This textbook gives a clear account of the manner in which knowledge in many branches of physics such as gravitation, thermodynamics, atomic physics and nuclear physics can be combined to gain an understanding of the structure and evolution of stars.

A major aim is to present the subject as one in which advances are still being made. The first half is an account of the observational properties of stars and a discussion of the equations that govern their structure. The second part discusses recent theoretical work on stellar evolution. The successes of the theory are stressed, but attention is also drawn to phenomena that are not completely understood.

This is a new edition of a widely-used textbook first published in 1970. New topics include mass loss from stars and close binary stars.

The Stars: their structure and evolution

The Stars: their structure and evolution

Roger J. Tayler

Professor of Astronomy
and Director of Astronomy Centre
University of Sussex

CAMBRIDGE
UNIVERSITY PRESS

Published by the Press Syndicate of the University of Cambridge
The Pitt Building, Trumpington Street, Cambridge CB2 1RP
40 West 20th Street, New York, NY 10011-4211, USA
10 Stamford Road, Oakleigh, Melbourne 3166, Australia

First Published by Wykeham Publications 1970
Second edition first published by Cambridge University Press 1994

Printed in Great Britain at the University Press, Cambridge

A catalogue record for this book is available from the British Library

Library of Congress cataloguing in publication data

Tayler, R. J. (Roger John)
The stars: their structure and evolution / Roger J. Tayler. —
2nd ed.
    p.   cm.
Includes index.
ISBN 0-521-46063-8. — ISBN 0-521-45885-4 (pbk.)
1. Stars—Structure. 2. Stars—Evolution. I. Title.
QB808.T37 1994
523.8—dc20   93-44924 CIP

ISBN 0 521 46063 8   hardback
ISBN 0 521 45885 4   paperpack

PN

Contents

Preface

This book is in effect a second edition of a book first published by Wykeham Publications in 1970. The Wykeham series was designed to bridge the gap between school and university science and the mathematical level of the book was designed to be suitable for sixth formers. In fact the main use of the book has been as a university textbook and some of the mathematics which was then taught in schools is now taught at university. In rewriting the book, I have not changed its general level, but I have introduced two appendices containing more mathematical detail relating to topics discussed in the main text.

Many branches of physics such as gravitation, thermodyamics, atomic physics and nuclear physics are combined in determining the structure of stars. As a result the subject provides an ideal example of the application of fundamental physics. Physical conditions in stars are more extreme than on Earth and a successful understanding of their structure should show how valid it is to extrapolate established physical laws to these conditions. Although it is profitable to study stars as isolated objects, an understanding of star formation and stellar evolution is central to the whole study of astronomy.

Significant progress has been made in explaining the observed properties of stars but there is still room for considerable improvement in the relation between theory and observation. In particular the process of star formation is not well-understood theoretically or observationally. In fundamental physics there is still need for a reliable theory of fully developed convection. A major aim of this book is to introduce the reader to work in a developing subject and the uncertainties of present theories are often emphasized.

Many workers have contributed to our present knowledge of stellar evolution but there are few names in the text, as it is impossible to apportion credit for every advance in the subject in a book of this size. Most of my diagrams are based on the results of other astronomers and my debt to them should be apparent.

I am grateful to Mrs Pauline Hinton for her very careful typing of the manuscript, with some help from Mrs Jöelle Nowers and Mrs Jane Walsh. I once again dedicate the book to my wife Moya, without whom this new version of the book would not have been completed.

R J Tayler

June 1993

Symbols

a	acceleration
A, Z, N	nuclear mass number, charge number, neutron number
B	magnetic induction
B_ν	Planck function
c_p	specific heat at constant pressure
d, d_*	distance, stellar
d	deuteron
$\mathrm{e}^-, \mathrm{e}^+$	electron, positron
E	energy
f	velocity distribution function
$f(M)$	initial mass function
$F_\mathrm{cond}, F_\mathrm{rad}$	conductive, radiative flux
g	acceleration due to gravity
h	angular momentum per unit mass
H_p	pressure scale height
i	inclination of orbit
I_ν	intensity of radiation
j_ν	emission coefficient
l	apparent luminosity, mixing length, mean free path
$L, L_\mathrm{s}, L_\mathrm{rad}, L_\mathrm{conv}, L_\mathrm{cond}$	luminosity, total, radiative, convective, conductive
m	apparent magnitude, fractional mass, particle mass
$M, M_\mathrm{V}, M_\mathrm{bol}$	absolute magnitude, visual, bolometric
$M, M_\mathrm{s}, M_\mathrm{crit}, M_\mathrm{cc}, M_\mathrm{ce}$	mass, total, critical, convective core, convective envelope
n	number density
n_f	number of degrees of freedom
n	neutron
O B A F G K M R N S	spectral types of stars

p	momentum
p	proton
$P, P_c, P_s, P_{gas}, P_{rad}$	pressure, central, surface, gas, radiation
P	period of orbit
q	charge on nucleus
Q	binding energy of nucleus
r, r_s	radius, total
R	Rayleigh number
R_{Sch}	Schwarzschild radius
t, t_d, t_{th}, t_n	time, dynamical, thermal, nuclear
$T, T_e, T_c, \overline{T}$	temperature, effective, central, mean
u	thermal energy per unit mass
U	total thermal energy
UBV	stellar magnitudes
$v, v_{esc}, \overline{v}$	velocity, escape, mean
v	volume per unit mass
V	volume
x	position
X, Y, Z	mass fraction of hydrogen, helium, heavy elements
α	helium 4 nucleus (alpha particle)
γ	ratio of specific heats
γ	photon
∇, ∇_{ad}	PdT/TdP, adiabatic value
$\varepsilon, \varepsilon_{pp}, \varepsilon_{CN}, \varepsilon_{3He}$	energy release, proton–proton, carbon–nitrogen, helium
η, λ, ν	exponents in laws of energy release and opacity
$\kappa, \kappa_s, \kappa_{rad}, \kappa_{cond}$	opacity, surface, radiative, conductive
κ_ν	absorption coefficient
λ	wavelength
$\lambda_{cond}, \lambda_{rad}$	thermal conductivity, radiative
μ	mean molecular weight
ν	frequency
$\nu_e, \overline{\nu}_e$	neutrino, antineutrino
$\rho, \rho_c, \overline{\rho}$	density, central, mean
σ_ν	scattering coefficient
Σ	surface density of disk
Φ	Roche potential
ω	angular velocity
ω_p	plasma frequency
Ω	gravitational potential energy

Numerical values

Fundamental physical constants

a	radiation density constant	$7.55 \times 10^{-16} \, \text{Jm}^{-3}\text{K}^{-4}$
c	velocity of light	$3.00 \times 10^{8} \, \text{ms}^{-1}$
G	gravitational constant	$6.67 \times 10^{-11} \, \text{Nm}^2 \, \text{kg}^{-2}$
h	Planck's constant	$6.62 \times 10^{-34} \, \text{Js}$
k	Boltzmann's constant	$1.38 \times 10^{-23} \, \text{JK}^{-1}$
m_{e}	mass of electron	$9.11 \times 10^{-31} \, \text{kg}$
m_{H}	mass of hydrogen atom	$1.67 \times 10^{-27} \, \text{kg}$
N_{A}	Avogadro's number	$6.02 \times 10^{23} \, \text{mol}^{-1}$
σ	Stefan–Boltzmann constant	$5.67 \times 10^{-8} \, \text{Wm}^{-2} \, \text{k}^{-4}$
$\mathscr{R}$	gas constant (k/m_{H})	$8.26 \times 10^{3} \, \text{JK}^{-1} \, \text{kg}^{-1}$

Astronomical Quantities

$L_{\odot}$	luminosity of Sun	$3.86 \times 10^{26} \, \text{W}$
$M_{\odot}$	mass of Sun	$1.99 \times 10^{30} \, \text{kg}$
$r_{\odot}$	radius of Sun	$6.96 \times 10^{8} \, \text{m}$
$T_{\text{e}\odot}$	effective temperature of Sun	$5780 \, \text{K}$
parsec	(unit of distance)	$3.09 \times 10^{16} \, \text{m}$

1

Introduction

This book is concerned with the structure and evolution of the stars, that is the life history of the stars. Its aim is to show how observations of the properties of stars and knowledge from many branches of physics have been combined, with the aid of the necessary mathematical techniques, to give us what we believe is a good understanding of the basis of this subject.

Because the stars are so remote from the Earth it may seem surprising that we can learn anything about their physical dimensions. To hope to be able to describe their internal structure and, still more, their evolution appears extremely optimistic. The mass and radius of a few stars can be measured directly, but for most stars the only source of information is in the light that we receive from them. This gives us some idea about the temperature and chemical composition of the *surface layers* of the star and about the total light output (*luminosity*) of those stars whose distance from the Earth is known. It also indicates that some stars are rotating rapidly or have strong magnetic fields and that others are losing mass from their surfaces. No direct information is obtained about physical conditions in the interiors of the stars, with the exceptions (discussed in Chapters 4 and 6) that the neutrinos emitted in the solar centre can be detected on Earth and that vibrations of the solar surface can provide information about the interior by techniques similar to seismology. The total of observational information which we have about the stars appears a small amount with which to hope to obtain an understanding of their internal structure.

If it seems presumptuous to hope to explain the present structure of stars, it is, perhaps, even worse when evolution is considered, for significant stellar evolution requires millions, or even thousands of millions, of years. Thus there are few instances of observation of stellar evolution and what there is can hardly be regarded as simple evolution. Some stars are observed to be losing mass into interstellar space, to be exchanging mass with a close companion or to be varying in their light output and occasionally a star explodes dramatically as a supernova, but there are no observations of changes in the properties of ordinary stars. For our nearest star, the Sun, there is no need of very detailed arguments to show that significant evolution must be very slow indeed. A small change in the Sun's properties would suffice to make the Earth uninhabitable for man, and man has been on Earth for hundreds of thousands, if not millions, of years. In fact, geologists say that the Earth's crust must have been solid for several thousand million years and that the Sun's luminosity cannot have changed significantly during that time. This gives an idea of the sort of time which is involved when we interest ourselves in the evolution of the Sun. I shall explain later that it is believed that more massive stars do evolve more rapidly, but even then we are usually concerned with periods of over a million years.

How then, is progress in this subject possible ? The main factor is that physics is a *relatively* simple subject with only a small number of fundamental laws. First, in the case of the structure of a star, we must be concerned with the forces which maintain it in equilibrium. At present it is believed that there are only four basic forces in nature (gravitational, electromagnetic, strong nuclear and weak nuclear) and only these can be involved in the structure of stars. The nuclear forces have a very short range and are not effective in holding together large bodies. The overall structure of a star is governed by the attractive force of gravity which pulls the star together and which is resisted by the thermal pressure of the material forming the star.

The main observational fact about stars is that they continuously radiate energy into space. This energy must have been released from some other source and have been transported from its point of release to the stellar surface. Perhaps the simplest idea would be to suppose that stars were created as very hot bodies and have been cooling down gradually ever since, but I shall show in Chapter 3 that it is impossible to reconcile this with the Sun's steady luminosity for such a long time. If this idea is discarded, the energy must have been converted into heat energy from another form inside the star, and it is then necessary to consider whether gravitational energy, chemical energy or nuclear energy might be involved. When the problem of the Sun's energy supply is considered in detail in Chapter 3, it becomes clear that only nuclear energy can meet the requirements and, in fact, essentially only one process, the conversion of hydrogen into helium with release of nuclear binding energy, can do it.

Of course, these facts that seem so obvious today were not always so clear. The structure and evolution of the stars were being studied before the properties of nuclear binding energy were fully understood and it was then thought that there must be a new unknown source of energy such as, perhaps, the *complete* annihilation of matter into radiation. At one time it was thought that the centres of stars were not hot enough for significant nuclear reactions to occur, and at this time Eddington made his famous suggestion that, if the centres of the stars were not hot enough, the nuclear physicists should look for a hotter place. I shall explain in Chapter 4 that the development of the quantum theory made this search unnecessary.

Although the basic forces of nature are few, the calculation of the structure of a star is not simple, as there are so many detailed physical processes which must be considered. It is necessary to have expressions for the rates of many nuclear reactions involved in the release of nuclear energy and even the conversion of hydrogen into helium involves several successive reactions. The energy released by the nuclear reactions must be carried from its point of release to the surface where it is radiated. I must therefore discuss whether the energy is carried principally by conduction, convection or radiation and must study the detailed processes involved in this transport of energy. As mentioned earlier, the pressure of the stellar material resists the attractive gravitational force tending to make a star smaller and the thermodynamic state of the stellar material must be studied so as to discover how pressure depends on temperature and density. In the discussion of the origin and transport of radiation and of the pressure of the stellar material, results will depend on the chemical composition of the star. Some information can be obtained about the chemical composition of the outer layers of a star from the occurrence of an element's characteristic spectral lines in the star's radiation, but it must be recognised that this might not be representative of the chemical composition of the star as a whole.

Because he has only limited information about the properties of *actual* stars, the theoretical astrophysicist tends to calculate the structure of a wide range of *possible* stars rather than trying to explain the properties of an individual star. According to the present theoretical ideas, a few basic properties of a star essentially determine its structure and evolution. The most important factors are believed to be mass and chemical composition, and calculations are made for a variety of different values for these. It then proves more useful to ask whether theory predicts a correct relationship between the properties of stars of different mass and chemical composition rather than whether it predicts the properties of an individual star, which are known only approximately. As I shall explain in the following paragraph, this procedure has been particularly useful because there are important regularities in the observed properties of stars. The only exception to this treatment of stars statistically rather then individually is that the Sun has received very detailed attention because we have so much information about it.

A major stimulus to the study of the evolution of the stars comes from the fact that, if one studies the value of mass, radius, luminosity and surface temperature for those stars for which values are available, it is found that not all combinations of values of these quantities are equally probable. The radius, luminosity and surface temperature are not independent because the energy radiated by unit area of the surface of a star is essentially determined by how hot it is. If I regard mass, luminosity and surface temperature as three independent quantities I can draw two independent diagrams relating them. It is usual to plot mass against luminosity (fig. 1) and luminosity against surface temperature (fig. 2) and in both of these diagrams most stars lie in quite narrow bands and there are large regions of the diagrams which contain no stars. For example, it is found that on the average the more massive stars are more luminous and have higher surface temperatures than less massive stars. One of the first tasks of stellar structure theory it to try to explain this regularity and it seems possible that there might be a reasonably simple explanation.

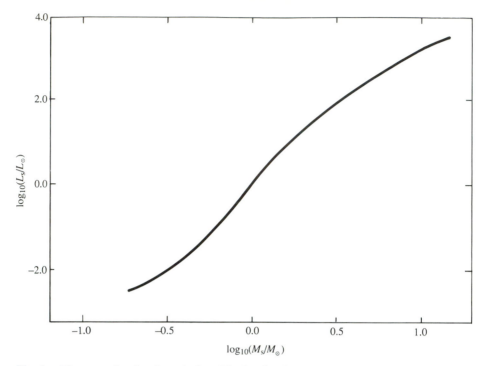

Fig. 1.   The mass–luminosity relation. The luminosity L_s is plotted against the mass M_s. $L_\odot$ and $M_\odot$ are the luminosity and mass of the Sun. Stars with accurately known luminosity and mass lie close to the curve shown, provided that they are main sequence stars (see Fig. 2).

It is believed that the three main factors determining the properties of a star are its *mass*, its *chemical composition* (when it is formed) and its *age*. Our observations of stars are complicated by their varying distances and the fact that obscuring matter produces an interstellar fog of varying density between us and the stars. Any attempt to interpret the properties of the whole group of stars for which there are good observational details is also complicated by the fact that the stars vary in mass, chemical composition and age. Such interpretation is easier for the groups of stars which are known as star clusters. These clusters of stars are apparently true physical groupings of stars rather than accidental concentrations, which happen to be in the same direction in the sky, but at very different distances. For a compact cluster it may be hypothesized that, of the five factors mentioned above which contribute to the appearance of a star, four, initial chemical composition, age, distance of the star from the Earth and the obscuring matter in the line of sight, might vary only slightly from star to star. If this is true, *the main factor which accounts for the differences in the observed properties of the stars is that they have different masses*. This has been the basis of most work on stellar evolution to date and it will be discussed in Chapter 6. Clearly, all of the five quantites do vary from star to star, but it seems reasonable that the variation of mass is most important.

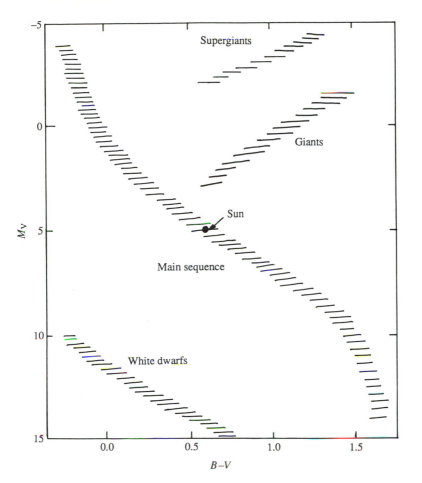

Fig. 2. The Hertzsprung–Russell diagram for nearby stars. The visual magnitude M_V is plotted against colour index $B-V$ and most stars fall in four well-defined groups. (M_V is proportional to $-\log L_S$ and $B-V$ is related to surface temperature, as shown in Table 1, page 17).

I have already mentioned that the Sun's properties are changing very slowly at present and it is further believed that slow variation of observational properties is characteristic of that phase in a star's evolution when nuclear reactions converting hydrogen into helium are occurring in the star's interior and are providing the energy that the star is radiating. This slow evolution prevents us from observing the rate at which stellar properties change, but it also has a very useful consequence in the theory of stellar structure. Because the hydrogen burning phase takes so long, the star settles down in a state which is almost independent of its previous history. This is useful because even now there is no very good theory of how stars are formed. If the study of stellar structure and evolution depended on having a comprehensive theory of star formation, the subject would have been much slower getting under way. Luckily it has been possible to regard the hydrogen burning phase of stellar evolution as the first stage.

Although the basic physical processes involved in stars, such as those concerned with energy release and energy transport, have been known since the 1930s and calculations of stellar structure had already been made even before all of the physical processes were understood, most of detailed work on stellar evolution has been done since 1960. This is largely because, when all of the physics of stellar interiors is to be taken into account, the equations of stellar structure and evolution can only be solved with the aid of a large computer and such computers did not become available until then. The advent of the large fast computer has revolutionized the amount of detail which can be included in the studies, although even now it is difficult to study very rapid stages of evolution particularly of non-spherical stars. This does not mean that there is no scope for less detailed calculations in which approximate values are used for some of the physical quantities in order to make the equations more tractable. In fact, as I shall discuss in Chapter 5, the general trends of luminosity and surface temperature as a function of mass can be understood on the basis of simplified physical laws. However, any detailed comparison between theory and observation requires the use of the most accurate mathematical expressions for physical laws.

It should be stressed that this is a book on a developing subject and that it is not an account of a field in which everything is understood. There are still some serious gaps in our knowledge and it has been my aim to mention and underline these rather than to pretend that they do not exist. Nevertheless I do feel that there is a good general understanding of the subject and believe, perhaps wrongly, that future changes will be ones of detail rather than upsets in the broad principles of the subject.

It is also hoped that what follows in the book will give some idea of how a research scientist approaches a problem. When a subject is completed it may be possible to present its development in a completely logical manner in which each step follows smoothly from the previous one. This is not, however, the situation when a subject is developing. Then it is more like working on a jigsaw puzzle. Pieces must be tried tentatively; the consequences of a variety of assumptions must be tried out. Parts of the subject may be studied in isolation in the hope that, when the whole pattern of the subject emerges, they will fit neatly into it. This is very much the situation with some parts of the subject of this book, particularly the contents of Chapters 7–10.

It should be clear from what has been said earlier that the study of stellar structure requires knowledge from many branches of physics such as atomic physics, nuclear physics, thermodynamics and gravitation. However, it should be stressed that the subject not only makes use of basic physical knowledge, but it also stimulates the development of further knowledge. In particular, I shall mention later that developments in nuclear physics have been stimulated by the need to understand the laws of energy release in stellar interiors and that study of final stages of evolution of massive stars is leading to interest in the behaviour of the law of gravitation in matter at extremely high densities. It should be remarked that the laws of physics as we understand them, have been obtained from experiments on the Earth and in its immediate environment. In studying the stars and the more distant parts of the Universe, *I make the assumption that the laws of physics are unchanging and are the same in all parts of the Universe*. This could be an incorrect assumption and, although I always try to understand astrophysical phenomena within the framework of existing physical laws, the possibility that this might be wrong must always be borne in mind.

Most of the book is concerned with the structure of isolated spherically symmetrical stars. This means, in particular, that mass loss from stars is largely ignored. In the past twenty years it has become apparent that mass loss is important at many stages of stellar evolution and that the final mass of a star may be very much less than its initial mass. In addition many of the more spectacular events in stellar evolution involve mass exchange between two stars which are close companions in a binary system. Thus departure from equilibrium and from spherical symmetry may be important in some stages of stellar evolution. At some stages in their evolution stars may suffer from instability which causes their properties to become variable. The detailed discussion of stellar stability and variable stars is too advanced for this book, but I shall explain the place in the evolutionary scheme of some of the important types of variable star.

I shall not discuss in any detail the importance of an understanding of stellar structure and evolution for astronomy in general. Stars are the most important component in the visible Universe and an understanding of galactic evolution requires a detailed knowledge of star formation, stellar evolution and mass loss from stars. Astronomers are now able to study galaxies at very large distances and hence in the remote past. The properties of such young galaxies are largely determined by the manner in which stars first formed in them.

The remainder of this book is arranged as follows. The observed properties of stars and the techniques of observation are described briefly in Chapter 2. The equations determining the structure of stars are discussed in Chapter 3. Included in these equations are quantities whose values can only be obtained by a rather detailed consideration of the physical state of stellar interiors, and the physics of stellar interiors is discussed in Chapter 4. The structure of hydrogen-burning stars at the beginning of their evolution, when nuclear reactions have just started to supply the energy the stars are radiating, is considered in Chapter 5. The chapter also contains a brief discussion of star formation and pre-main-sequence evolution. The early evolution of these stars is discussed in Chapter 6. Mass loss from stars is discussed in Chapter 7 and the properties of close binary stars in Chapter 8. Chapters 9 and 10 are concerned with later stages of stellar evolution. Finally, some of the problems for future study are described in Chapter 11.

2

Observational properties of stars

Introduction

The subject matter of this book is stars and in particular the properties of individual stars but, before I start discussing these properties, I give a general description of the Universe in which the stars are situated and of which they may be the most important component. The *may* in this sentence be very important. At one time there would have been little doubt that stars are the most important constituent in the Universe. More recently it has become clear that there may be a considerable amount of material in the Universe which is not in the form of stars and it is possible that most of the mass in the Universe is in the form of weakly interacting elementary particles. In giving a brief description of the Universe, no attempt will be made to explain how the results are obtained, but subsequently a detailed discussion will be given of how the properties of stars are deduced from observation.

With the naked eye on a clear night one can observe a few thousand stars and it can be seen that there is a region in the sky, known as the Milky Way, in which there is a particularly large density of faint stars. With even a small telescope, the number of stars which can be seen is greatly increased and it is now known that the solar system belongs to a large flattened system of stars known as *the Galaxy*, which probably contains about 100 000 million stars. Schematic views of the Galaxy as it would look from outside are shown in figs. 3 and 4. The main bulk of the stars in the Galaxy are contained in a highly flattened disk with a central bulge (nucleus), although there are stars in smaller numbers throughout an approximately spherical halo. The diameter of the disk is of the order of 100 000 light years, where a light year is the distance travelled by light in one year (9.5 x 10^{15} m).† The thickness of the disk is only about 1000 light years so that it can be seen that it is very highly flattened indeed.

---

† This unit is not usually used by professional astronomers who use the parsec defined on page 18.

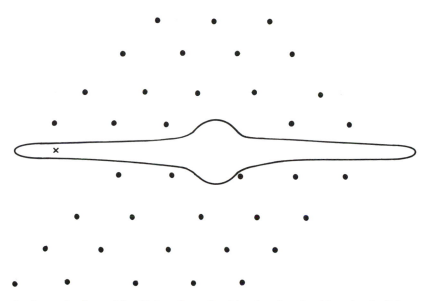

Fig. 3. A schematic view of the Galaxy from the side, showing the thin galactic disk and the central nuclear bulge. The position of the Sun is marked with a cross and the filled circles represent globular clusters.

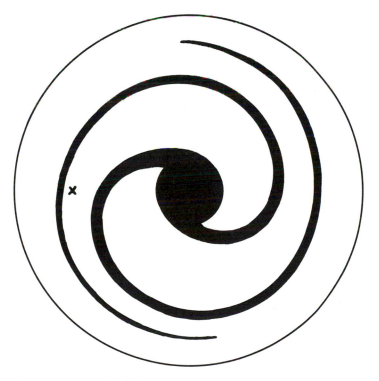

Fig. 4. A schematic view of the Galaxy from above. The spiral structure is shown and the position of the Sun is marked with a cross.

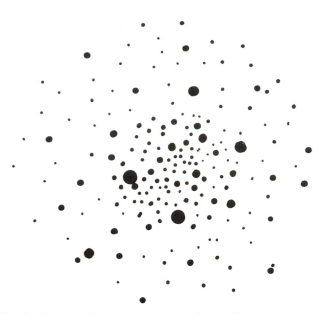

Fig. 5.   The distribution of stars in a globular cluster. The brightest stars are shown as they would appear in a photograph.

Many of the stars in the Galaxy are members of binary or multiple systems. A binary system contains two stars which are held together by their mutual gravitational attraction and which describe orbits about their centre of mass. Multiple systems are larger groups of stars which are held together by their mutual gravitational attraction. We shall see below that much of our detailed knowledge about stars is obtained from a careful study of binary systems. Of course, all the stars in the Galaxy move under the gravitational attraction of all the other stars, but a star which has no very close neighbours moves more or less freely and in a straight line for quite long periods of time. Many stars are also members of large sub-systems known as star clusters and we shall find later that the existence of these star clusters is very important for the subject matter of this book. Clusters are of two general types, globular and galactic (or open), although there is no completely clear division between the two types. The general appearance of globular and galactic clusters is shown in figs. 5 and 6. Globular clusters have a compact circular appearance, they are spread throughout the Galaxy including the halo, there are more than 100 of them and they contain between 100 000 and 1 000 000 stars each. The galactic clusters contain many fewer stars. They are called galactic because they are situated in the plane of the Galaxy and open because their appearance is diffuse rather than compact. There are several hundred galactic clusters known.

As shown in fig. 4, the disk of the Galaxy has a spiral structure. Many of the brightest stars in the Galaxy are found in, or near, the spiral arms. As well as stars the Galaxy contains clouds of gas and dust. The gas is also largely situated in the

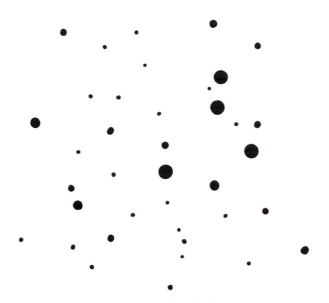

Fig. 6. The distribution of stars in a galactic cluster.

spiral arms and it may account for between 1/20 and 1/10 of the mass of the Galaxy. The interstellar gas is believed to be the material out of which stars are formed. We shall see later that the bright stars in the spiral arms are thought to be stars which have been formed quite recently in the galactic lifetime, and it then seems quite natural that they are closely associated with the interstellar matter out of which stars are still forming today.

At one time it was thought that the Galaxy was the whole Universe, although there were objects called spiral and elliptical nebulae whose position in the Galaxy was unclear. It is now known that these nebulae are also galaxies and, in particular, some of the spiral nebulae are very similar to our Galaxy. Galaxies have been observed out to distances of a few thousand million light years, and with galaxies being typically a few million light years apart, this means that there are many thousands of millions of galaxies in the observable Universe. The stars in these galaxies appear to be similar to the stars in our own Galaxy and the theory of stellar structure developed in this book should be applicable to all of them, although it is only in our immediate neighbourhood in our Galaxy that stellar properties can be observed in fine detail. A discussion of the properties of all the galaxies in the Universe would soon involve us in a discussion of cosmological theories. These theories are concerned with whether the Universe had an origin in time or whether it has always existed, whether all the galaxies were formed at about the same time or whether galaxies are still being formed today and many similar questions. I do not discuss cosmology in this book; in fact, the study of the life histories of stars in our own Galaxy is essentially independent of wider cosmological questions. The properties of galaxies are discussed in a companion book *Galaxies: structure and evolution*.

I shall now return to a consideration of the properties of individual stars. We shall find there are many gaps in our knowledge, but that nevertheless a reasonably consistent picture emerges.

Luminosity, colour and surface temperature

Most information about stars is obtained from the light and the other electromagnetic radiation which they emit. Observations give us some information about both the *quantity* and the *quality* of this light. In principle we can detect the amount of radiation from a star which falls on unit area of the Earth's surface and we can investigate what is the distribution with wavelength of the radiation. There are many different detection systems used. These include direct photography with a photographic plate, which is sensitive to a rather wide wavelength range, and the use of prisms or diffraction gratings to spread the light out into a spectrum before it falls onto the photographic plate. There are also many devices based on the photoelectric effect which detect the electrons emitted from a *light sensitive surface*. Those most commonly used today are known as charge coupled devices or CCDs. In most cases filters are used to cut out all of the light except in a narrow wavelength range and this is known as *narrow band photoelectric photometry*. If the entire energy output of a star, irrespective of its wavelength, is to be measured, a *bolometer* or *pyrometer* may be used; these measure the energy received in the form of heat.

For some purposes it is useful or essential to detect the radiation in narrow wavelength bands, while for others it is more useful to have a detector with as wide a wavelength response as possible. At present the theoretical astrophysicist finds it much easier to predict the total output of radiation from a star of given mass and chemical composition than to calculate its exact distribution over wavelength. Thus for comparison with theory it is desirable to measure the radiation over as wide a range of wavelength as possible either by the use of many narrow wavelength band detectors spread over the whole spectrum, or by the use of a bolometer which responds to the entire wavelength range of interest. In what follows I shall refer particularly to two types of observation. These are spectroscopic observations which are essential for discussions of the chemical composition of stars and the photoelectric measurements in the three wavelength bands known as U, B and V which are centred in the ultraviolet, blue and yellow regions of the spectrum and which will be defined and discussed further below.

Magnitudes

Observations of the light received from a star are normally expressed in *magnitudes*. Magnitude is a logarithmic measure of luminosity with the brightest stars having the lowest magnitude. This convention arose because the Greek astronomers originally catalogued the stars visible to the naked eye in six magnitudes, with the first magnitude stars being the brightest. When, in the

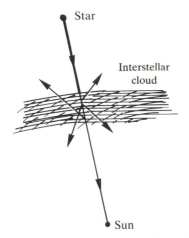

Star

Interstellar
cloud

Sun

Fig. 7.   Scattering and absorption of starlight by an interstellar cloud.

nineteenth century, a quantitative system was first introduced, it was made to agree as closely as possible with the old qualitative measures. Thus

$$m = \text{const} - 2.5 \log l_s, \tag{2.1}$$

where m is the magnitude, l_s is the luminosity (total light energy received) in the wavelength range being studied and the constant is used to define the zero of the magnitude scale. Such a magnitude scale, with the zero point chosen appropriately, gives reasonable agreement with the earlier estimates because the human eye most closely responds to the logarithm of the luminosity than to the luminosity itself.

Note that the luminosity entering into (2.1) is the apparent luminosity, in the sense that it refers to the amount of radiation which falls on unit area of a detector on the Earth's surface. If I am to discuss the intrinsic properties of stars, I must convert this to the absolute luminosity of the star, the energy emitted by the star per second. Note that apparent luminosity and absolute luminosity have different dimensions, but this should not cause confusion. In order to convert apparent luminosity to absolute luminosity I first need to know the distance of the star from the Earth. Then the amount of radiation falling on the unit area of the Earth's surface perpendicular to the direction to the star can be multiplied by $4\pi d_*^2$, where d_*^2 is the distance of the star from the Earth. This alone is difficult because there are not many stars whose distance can be measured directly (see the discussion later in this chapter).

However, the problem is more difficult than this because radiation can be scattered or absorbed by material between us and the star (see fig. 7) either by the gas in interstellar space or by the Earth's atmosphere. Until quite recently astronomical observations have been restricted to the wavelength ranges in which there is an atmospheric *window* (see fig. 8). It is at first sight fortunate, but presumably in no way accidental, that the visible *window* almost coincides with the wavelength range to which the human eye is sensitive and that it is a range

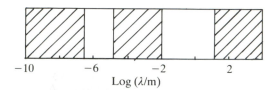

Fig. 8.   The transparency of the Earth's atmosphere. Electromagnetic waves of wavelengths corresponding to the hatched areas are almost completely absorbed in the Earth's atmosphere. Between the hatched areas are the visible *window* and the radio *window*.

which contains most of the radiation emitted by the Sun and many other stars. With the advent of rockets and artificial satellites and the placing of telescopes in space above the Earth's atmosphere, this difficulty has been at least partially removed. The same is not true of the effect of interstellar matter between us and the stars and, although it is possible to estimate its influence, some uncertainities remain.

I can write down an equation which relates the quantity of radiation emitted by the star to the amount observed. Suppose first that $L_\lambda d\lambda$ is the total energy emitted by the star in the wavelength range between λ and $\lambda + d\lambda$, so that the luminosity (total rate of emission of energy) of the star is

$$L_s = \int_0^\infty L_\lambda \, d\lambda. \tag{2.2}$$

If the Earth's atmosphere and interstellar space were transparent to radiation, the energy reaching unit area of the Earth's surface perpendicular to the direction to the star, per second in wavelength range $d\lambda$ would be $L_\lambda d\lambda/4\pi d_*^2$. I can introduce a quantity t_λ which is the probability that radiation of wavelength λ will reach the Earth's surface and a second quantity s_λ which measures the sensitivity of the detection system used. Then the amount of energy detected per unit area at the Earth's surface in wavelength range $d\lambda$ around the wavelength λ is :

$$l_\lambda d\lambda = L_\lambda d\lambda t_\lambda s_\lambda/4\pi d_*^2, \tag{2.3}$$

and the total energy detected is :

$$l_s = \int_0^\infty (L_\lambda t_\lambda s_\lambda/4\pi d_*^2) \, d\lambda. \tag{2.4}$$

Surface temperature

By the *quality* of the radiation emitted by a star is meant its distribution in terms of wavelength or alternatively frequency.† The crudest measure of the quality of the light is its colour and this is often referred to in descriptions of stars

---

† In the remainder of this book I shall describe light by frequency rather than wavelength. The two are, of course, related by $\lambda\nu = c$, where ν is the frequency and c the velocity of light.

as red giants, white dwarfs, blue supergiants, etc. A complete description of the quality of the light involves the measurement of l_λ at all wavelengths. The quality of the light which we receive from a star is not affected directly by the distance of the star from us; the quantity of radiation is reduced by the same geometrical factor $4\pi d_*^2$, at all wavelengths. The quality of radiation is affected by the Doppler effect if the star being studied is moving towards or away from us. Although this Doppler effect which shifts spectral lines either towards the red or the blue can be used to deduce the velocity of the star, it only has a significant effect on the quality of the radiation if the velocity is comparable with the velocity of light. This is true of some distant galaxies which are receding from us with high velocities, but not of stars in our own and nearby galaxies. However, as is clear from (2.3), the quality of the light is affected by absorption and scattering which do not act equally on all wavelengths and this fact helps us to unravel how much absorption has occurred.

The colour of a star is related to its surface temperature, although the latter cannot be uniquely defined. A temperature can be defined when a system is in a state of *thermodynamic equilibrium*† and the the distribution of radiation with frequency is uniquely defined by the temperature and follows the *black body* or Planck law. In these circumstances the amount of radiant energy crossing unit area, in a unit solid angle about the direction normal to the area, in unit frequency range and in unit time is $B_\nu(T)$, where

$$B_\nu(T) = \frac{2h\nu^3}{c^2} \frac{1}{\exp(h\nu/kT) - 1}. \tag{2.5}$$

In (2.5), B_ν is called the Planck distribution at temperature T K, ν is the frequency, c is the velocity of light (3×10^8 m s^{-1}), h is Planck's constant (6.6×10^{-34} J s) and k is Boltzmann's constant (1.4×10^{-23} J K^{-1}). In practice we often use the word temperature when a state of thermodynamic equilibrium does not exist; in particular we often use it as a measure of the mean kinetic energy of the particles present. For the stars we do not have very direct information about the particles and we must try to deduce a surface temperature from the radiation which we receive. Planck curves for three values of T are shown in fig. 9.

For some stars the distribution of energy with frequency is not too different from the black body curve (2.5) and for these stars a surface temperature can readily be defined. For others this is more difficult and observers now usually use a less subjective measure of the quality of the light known as colour index instead of surface temperature. A colour index is the difference between the magnitude of the star in two wavelength bands. Thus, if we use the three colour U, B, V photoelectric system mentioned earlier, we can define three colour indices U–B, B–V, U–V, where the symbol U, for example, is now being used to denote the magnitude of the star in the U band. The filters used for each of these bands admit radiation in a range of wavelength about 1000 Å wide and their central wavelengths are approximately:

$$\lambda_U \approx 3650 \text{ Å}, \qquad \lambda_B \approx 4400 \text{ Å}, \qquad \lambda_V \approx 5480 \text{ Å}. \tag{2.6}$$

† The concept of thermodynamic equilibrium is discussed in Appendix 1.

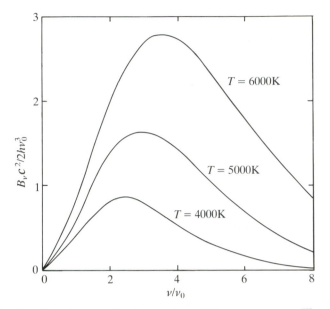

Fig. 9.   Planck curves for three values of the temperature. The normalising frequency ν_0 is 10^{16}s^{-1}.

Here $1\,\text{Å} \approx 10^{-10}$ m is a unit used by astronomical spectroscopists. If a star did radiate as a black body the colour index would be directly related to the logarithm of the surface temperature; in general it is an approximate measure of surface temperature which is less subjective than estimated surface temperatures.

Effective temperature and bolometric correction

Later in this book I shall be concerned with making a comparison between observed properties of stars and those predicted by solutions of the equations of stellar structure. As mentioned earlier the theoretical astrophysicist finds it much easier to predict the total amount of energy radiated by a star than to calculate its distribution with frequency. Theoreticians define what they call the effective temperature of a star, T_e. This is defined in such a way that a black body of temperature T_e with the same radius as the star would radiate the same total amount of energy. Thus

$$L_s = \pi a c r_s^2 T_e^4 \equiv 4\pi r_s^2 \sigma T_e^4, \tag{2.7}$$

where r_s is the radius of the star and a is the radiation density constant (7.55×10^{-16} J m^{-3} K^{-4}) and σ ($\equiv ac/4$) is the Stefan–Boltzmann constant (5.67×10^{-8} W m^{-2} K^{-4}).[†] From the second equality in (2.7) it can be seen that the energy radiated by unit area of a black body of temperature T_e is σT_e^4.

---

† There is considerable confusion in the astronomical literature concerning the naming of the two constants a and σ. In particular in many books concerned with the structure of stars, a is called the Stefan–Boltzmann constant while in many other books σ is called the Stefan–Boltzmann constant. Our present usage is now standard.

Table 1. *The relationship between colour index, B–V,*
absolute visual magnitude, M_V, logarithm of effective
temperature and absolute bolometric magnitude, M_{Bol},
for main sequence stars

B–V	M_V	$\log T_e$	M_{Bol}	B–V	M_V	$\log T_e$	M_{Bol}
−0.3	−4.4	4.48	−7.6	0.5	3.8	3.80	3.8
−0.2	−1.6	4.27	−3.5	0.6	4.4	3.77	4.3
−0.1	0.1	4.14	−0.8	0.7	5.2	3.74	5.1
0.0	0.8	4.03	0.4	0.8	5.8	3.72	5.6
0.1	1.5	3.97	1.3	0.9	6.2	3.69	5.9
0.2	2.0	3.91	1.9	1.0	6.6	3.65	6.2
0.3	2.3	3.87	2.2	1.1	6.9	3.62	6.4
0.4	2.8	3.84	2.8	1.2	7.3	3.59	6.6

One of the main problems in correlating theories and observations of stars is that of relating effective temperature to colour index or other estimates of stellar surface temperature and bolometric luminosity to magnitude measured in some particular wavelength range. Thus, in particular, we are often interested in the transformation from (L_s, T_e) to $(V, B–V)$. Ultimately such a transformation can only be made by measuring the total energy output of the star with a bolometer or by measuring magnitudes in a large number of narrow wavelength bands. In practice such observations can only be made for a limited number of stars, but these can be used to set up an empirical relation between effective temperature and colour index and between bolometric magnitude and visual magnitude which can then be applied to other stars. Such transformations for main sequence stars (the term will be defined on page 33) are shown in Table 1.

Absolute magnitude

The definition of magnitude given in (2.1) is in terms of the amount of radiation received on unit area at the Earth's surface and it is known as the *apparent magnitude* of the star. We often wish to use a magnitude which is a measure of the total light emitted by the star. The *absolute magnitude* of a star is defined to be the apparent magnitude it would have if it were placed at a distance of 10 parsecs, where the parsec will be defined in the next paragraph. If the actual distance of a star is d parsecs, its absolute magnitude M and apparent magnitude m are related by:

$$M = m - 5 \log (d/10). \tag{2.8}$$

Stellar distance

In order to convert apparent magnitude into absolute magnitude, the distance of the star is required. This can be obtained directly for a relatively small

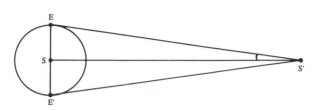

Fig. 10.   The parallactic angle of a star.

number of nearby stars. For these it can be measured by a trigonometric method. Suppose I consider the Earth's motion around the Sun (fig. 10). The apparent direction of a star, measured relative to the position of much more distant stars, changes as the Earth describes its orbit around the Sun. If this angular displacement can be measured, the triangle EE'S' can be solved to find the distance of the Earth (or Sun) from the star. The angle ES'S is called the *parallactic angle* of the star. For the nearest star (other than the Sun) the angle is somewhat less than 1″, which means that the distance to the nearest star is very great indeed. Because of the size of astronomical distances, astronomers use a distance scale based on the distance at which the parallax would be 1″ known as the parsec, where

$$1 \text{ parsec} = 3.09 \times 10^{16} \text{m}, \tag{2.9}$$

or

$$1 \text{ parsec} = 3.26 \text{ light years}. \tag{2.10}$$

The discussion of the measurement of stellar distance given above has been rather simplified. Thus it has been assumed that the star being observed is at rest relative to the Sun. In fact stars in the Galaxy are not at rest, but they are describing orbits in the mutual gravitational fields of all the other stars with velocities relative to one another of the order of 10^4 to a few times 10^5ms^{-1}. If the star being studied moves in the direction perpendicular to the line SS' a distance comparable to, or greater than, the distance EE' in the six months that it takes the Earth to travel from E to E', as it can with the velocities quoted above, the method described will not give the correct answer (the distance moved in the line of sight is unimportant because even for the nearest stars this is minute compared with the distance SS'). However, the motions of stars are such that, unless they are partners in close binary systems, they can be assumed to move with a uniform velocity in a straight line for periods measured in years, and deviations will only become apparent after thousands or millions of years. This means that, by observations over a period of several years, the steady displacement of the star with respect to very distant stars, which is produced by its own motion, can be separated from its periodic displacement due to the Earth's motion about the Sun (fig. 11).

As the largest parallax for any star is less than 1″, it is perhaps surprising that any parallaxes can be measured. In fact the parallaxes of three stars, 61 Cygni, α Lyrae

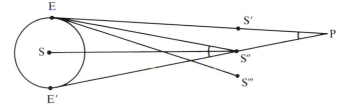

Fig. 11.   The parallax of a moving star. As the Earth moves from E to E' and back to E, the star moves from S' to S" and S'". For a distant star, the true parallactic angle is ES"S which cannot be directly observed. The angles EPE' and S'ES" can be observed and, assuming that the star is moving with constant velocity, the parallax can be obtained by simple geometry.

and α Centauri†, were first measured by three different observers in the year 1838. Parallaxes can be measured with some degree of accuracy down to about 1/50" (distances of 50 parsecs) and they are known for several thousand stars. In terms of the dimensions of the Galaxy which have been mentioned earlier in the chapter, 50 parsecs is a very small distance and it is clear that direct distance measurements can only be made for the nearest stars. For more distant stars indirect methods of estimating distance must be used and some of these will be mentioned later in this chapter.

Proper motion

In measuring distance I have separated out the periodic displacement of a star caused by the Earth's motion from its regular displacement produced by it own motion. This apparent angular motion of stars across the sky perpendicular to the line of sight is called the *proper motion*. Measurements of proper motion give some information about the way in which stars are moving, but the apparent motion cannot be converted into true velocity unless the distance of the star is known (see fig. 12). However, observations of proper motions are useful in helping us to discover nearby stars whose distance can be measured. Parallaxes can be measured for nearby stars, but these stars do not carry labels saying *nearby star*. There are two obvious way of trying to identify nearby stars. The first is to assume that many of the stars which appear very bright are also very near to us. The other is to choose stars with large proper motion. Unless there is a steady increase in velocity of stars with distance from the Sun, which seems unlikely as the Sun is not at the galactic centre, it is likely that stars which have large proper motions are also nearby stars. Thus the first stars to study for parallaxes are apparently bright stars with large proper motions.

† Stars are named by referring them to the constellations in which they occur. The brightest stars are denoted by Greek letters and fainter stars by numbers. Thus α Lyrae (Vega) is the brightest star in the constellation Lyra. The constellations are apparent groupings of stars in the sky in contrast to clusters which are physical groupings.

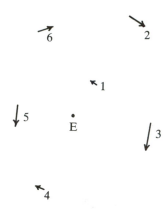

Fig. 12.　The motions of six stars in a given period of time are indicated by the lengths of the arrows. Their *proper motion* in that time is the angle subtended by the arrow at the Earth, E. Stars 1 and 2 have different true motions but identical proper motions.

The study of parallaxes, proper motions and radial velocities (velocities in the line of sight deduced from the Doppler effect on spectral lines) give us information about the positions and motions of stars and hence about the structure of our neighbourhood in the Galaxy. This, however, is not the subject matter of the present book. It is covered in the companion book, *Galaxies: structure and evolution*.

Stellar masses

There is only one direct way of obtaining a stellar mass and that is by studying the dynamics of a binary system. The method used depends on whether the binary system is a wide binary or a close binary. According to Kepler's laws, the two stars in a physical binary system (that is a genuine binary system rather than two stars which are in the same direction from us but which are at totally different distances) revolve about their centre of mass in elliptic orbits (fig. 13). If the stars in a binary are sufficiently far apart, in relation to their distance from the Earth, both stars can be observed and over a sufficiently long period of time their orbits about their centre of mass can be studied. If the parallax of the binary system can be measured so that its distance from the Earth is known, the apparent size of the orbits can be converted to a true size. This size of the orbit combined with the period of revolution of the stars enables the masses of both stars to be determined by use of Kepler's laws. This is discussed on the next page.

In fact the problem is not quite as simple as that. In fig. 13 the orbits of the stars have been drawn as they would appear if the orbital plane were perpendicular to the line of sight from the Earth. For most systems this will not be true. If sufficiently precise observations of the motion of the stars could be made for a long enough period of time, it would become clear that the centre of mass was not at the focus of the apparent orbits and the inclination of the orbital plane to the line of

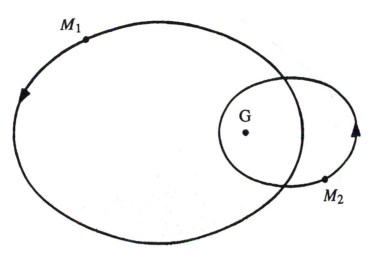

Fig. 13.   The elliptical orbits of two stars of masses M_1 and M_2 about their centre of mass G.

sight could be deduced from the fact that it must be at the focus of the true orbits; in practice this reduction is likely to be difficult. The problem is simplified if the eccentricities of the ellipses are small because, however a circular orbit is tilted, the maximum apparent dimension of the orbit is its diameter. In addition, if orbits appear almost circular, they must be approximately at right-angles to the line of sight.

Suppose that a double star system has a known parallax so that its distance from us is known and that in addition the two stars are observed to move in circular orbits (fig. 14). As the distance of the system is known, the apparent size of the orbit can be converted into the real size so that the radii r_1, r_2 of the two orbits are known. The centre of the orbits is the centre of mass of the system so that $M_1 r_1 = M_2 r_2$ or

$$M_1/M_2 = r_2/r_1, \tag{2.11}$$

where M_1 and M_2 are the masses of the two stars. Thus, as r_2 and r_1 are known, so is the ratio of the masses. Another relation between the masses can be obtained from Kepler's laws. These give a relation between the distance between M_1 and M_2, the sum of the masses and period of revolution P (the time taken for either star to describe its orbit). This relation is:

$$P^2 = 4\pi^2 (r_1 + r_2)^3 / G(M_1 + M_2). \tag{2.12}$$

Equations (2.11) and (2.12) enable the masses of both stars to be determined.

Clearly, if the parallax of the binary system cannot be measured, only the apparent dimensions of the orbit will be known and in that case the masses of the individual stars cannot be found. However, the ratio of the masses will be known from (2.11). If M_2 is very much more massive than M_1, it will probably be

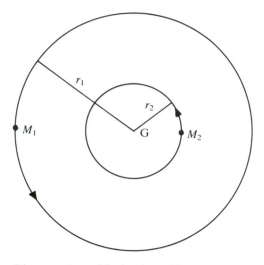

Fig. 14.   Binary system with circular orbits.

impossible to determine r_2 accurately, even if the parallax can be measured. In that case (2.12) becomes approximately:

$$P^2 = 4\pi^2 r_1^3/GM_2 \qquad (2.13)$$

and the mass of the principal component can be obtained.

Eclipsing binaries

It is also possible to obtain the masses of some close binary systems, but here the technique used is quite different. It involves the study of the spectra of the two components of the system. Before I discuss mass determination from the study of *spectroscopic binaries* I will first discuss *eclipsing binaries*. Such a binary is most simply described if it can be assumed that the less massive component moves in a circular orbit about the more massive component. If the plane of the orbit of a close binary system is almost such that it contains the line of sight from the Earth, it is possible that one star is eclipsed by the other at some stage during the orbit (fig. 15). The effect of this eclipse is to produce an apparent variation of the light output of the binary system as is shown in fig. 16. Quite a large amount of information can be obtained from such a light curve. In the first place the period of rotation of the secondary star around the primary can be found. The duration of the eclipses compared to the time between successive eclipses gives information about the radii of the stars compared to the size of the orbit. Finally, from the depth of the eclipses it may be possible to learn something about the angle of inclination of the plane of the orbit to the line of sight; the angle of inclination, i, is defined as the angle between the line of sight and the perpendicular to the orbit plane (fig. 17) and for eclipsing binaries this must be quite close to $90°$.

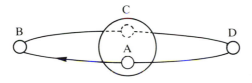

Fig. 15. Eclipsing binary. In position A the smaller star eclipses part of the larger star; at position C it is eclipsed.

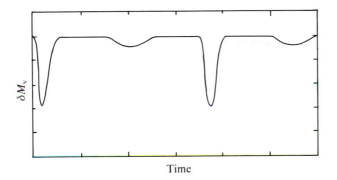

Fig. 16. The light curve of an eclipsing binary. Such a light curve would be obtained for the system of fig. 15 if the smaller star were very much hotter and the deep minimum would correspond to the eclipse of the hotter star.

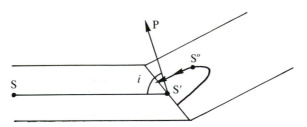

Fig. 17. The angle of inclination of a binary system. S represents the Sun and S′ and S″ the stars of a binary system, S′ being assumed (for simplicity) to be much more massive that S″. S′P is perpendicular to the orbital plane and the angle i, SS′P, is called the angle of inclination.

Spectroscopic binaries

If the light from a close binary system is studied it may be possible to separate the spectral lines of the two components of the system. When the stars are at a phase in their orbits when one star has a component of velocity towards the Earth and the other has a component of velocity away from the Earth, the spectral lines of the two stars are separated by the Doppler effect and it is possible to obtain the velocities of the two stars. If the angle of inclination were 90°, we would be able to observe the actual velocities of the two stars, but for an arbitrary angle of inclination only $v_1 \sin i$ and $v_2 \sin i$ can be observed, where v_1 and v_2 are the velocities of the two stars.

As each star takes the same time to describe the orbit, the velocity of either star is proportional to the radius of its orbit. In addition, from (2.11), the velocity is inversely proportional to the mass. Thus:

$$\frac{v_1 \sin i}{v_2 \sin i} \equiv \frac{v_1}{v_2} = \frac{r_1}{r_2} = \frac{M_2}{M_1}. \tag{2.14}$$

Thus the observation of $v_1 \sin i$ and $v_2 \sin i$ immediately give a value for the mass ratio of the binary system. If v_1 and v_2 were known, we could obtain the radii of the orbits (assumed circular) from the observed period through the relations:

$$v_1 P = 2\pi r_1, \qquad v_2 P = 2\pi r_2. \tag{2.15}$$

However, as only $v_1 \sin i$ can be observed, only values of $r_1 \sin i$ and $r_2 \sin i$ can be obtained. Equation (2.12) can then be rewritten :

$$(M_1 + M_2) \sin^3 i = 4\pi^2 (r_1 + r_2)^3 \sin^3 i / GP^2, \tag{2.16}$$

where all the quantities on the right-hand side of (2.16) can be obtained from the observations. As spectroscopic binaries can be observed with arbitrary values of the angle of inclination, only a lower limit of the sum of the masses can be obtained from (2.16):

$$M_1 + M_2 \geq 4\pi^2 (r_1 + r_2)^3 \sin^3 i / GP^2. \tag{2.17}$$

If a spectroscopic binary is also an eclipsing binary further progress can be made. In the first place, as the angle of inclination must be very close to 90° for the system to have eclipses, (2.17) can be replaced by:

$$M_1 + M_2 \approx 4\pi^2 (r_1 + r_2)^3 \sin^3 i / GP^2, \tag{2.18}$$

and the masses of the two stars can be found from (2.14) and (2.18). In addition the true dimensions of the orbit are known from (2.15). The comparison of the time taken for eclipses with the period P now enables values to be obtained for the radii of the stars and this is one of the few ways in which stellar radii can be estimated. I now consider the problem of the measurement of radii.

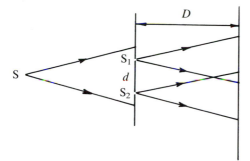

Fig. 18.   Interferometry. Light from a source S falls on a screen containing two slits S_1 and S_2 and a pattern of bright and dark lines is formed on a second screen.

Stellar radii

There are three different ways of obtaining values for the radii of stars. These involve direct measurements of stellar angular diameters by interferometry, study of eclipses as described above and use of the relation (2.7) between luminosity, radius and effective temperature.

The simplest interferometric method can be described as follows. If light from a point source is allowed to fall on a screen containing two slits (fig. 18) and the light from the slits is then allowed to fall on another screen, a pattern of bright and dark lines is observed on this second screen. This is the phenomenon known as interference, which can easily be understood on the wave theory of light and which was one of the main experimental facts leading to the development of that theory. The distance between successive maxima and minima in the intensity is :

$$x = D\lambda/d, \tag{2.19}$$

where λ is the wavelength of the light, D is the distance between the two screens and d is the distance between the slits.

If the original source of the light is not a point source, or if the slits are too wide, the interference pattern may be destroyed. Thus, if the angular diameter of the source at the slit is θ (fig. 19), the interference fringes produced by light from different parts of the source overlap as shown in the figure and the interference fringes disappear when the distance between the slits is larger than

$$d = A\lambda/\theta, \tag{2.20}$$

where A is a number which is of the order of unity and which depends on the shape and density of illumination of the source. For a uniformly bright circular disk $A = 1.22$ and A exceeds 1.22 for a disk which is darker at the edges, as the solar disk is observed to be. This increase of A for sources with their emission concentrated towards the centre is easy to understand. Such a source really behaves as a source of smaller angular diameter; from (2.20) reducing θ and increasing A give similar results.

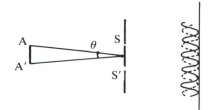

Fig. 19.   If the source of fig. 18 is extended with diameter AA′, light from different parts of the source produces out of phase intensity patterns at the second screen as shown by the solid and dashed lines. If these patterns are sufficiently displaced, the overall interference pattern disappears.

Most stars have such small angular diameters that with practicable slit separations the interference pattern is observed and they appear as point sources. Some do have large enough angular diameters that it is possible to find the value of d at which the interference pattern disappears and that leads to an estimate of the angular diameter of the star provided a value can be obtained for A. Theoretical values of A are available for stars of different types. Once the angular diameter of a star is known, its linear diameter can be obtained provided that its distance is known. Clearly this method will preferentially provide diameters of nearby stars of large diameter.

The estimation of radius using the properties of eclipsing binaries has been briefly described above. One point should be added. At the time the eclipse is occurring, the star is moving transverse to the line of sight, apart from any motion of the binary system as a whole towards or away from the Sun. At two other times in the star's motion it is moving either towards or away from the Sun and the difference in these velocities should enable the true orbital velocity of the star to be obtained. The time taken for the eclipse combined with this velocity then gives the radius of the star and this measurement does not require knowledge of the distance of the binary system.

The method of eclipses can be used in other ways. For example, a star may be in such a position in the sky that it apparently passes behind the Moon; this is known as lunar occultation. In principle the time taken for the star's light to disappear from the moment of first contact gives a value of its angular diameter because the angular speed of the Moon is well known. This diameter can be converted into a true diameter if the star's distance is known. In actual fact the method of lunar occultation is not as simple as this because of diffraction of the star's light as it passes behind the Moon. However, the angular diameter of the star can be obtained from the properties of the diffraction pattern. Interferometric techniques and the method of lunar occultations have also been much used by radio astronomers in their study of angular diameters of cosmic radio sources.

Finally, radius can be estimated using (2.7)

$$L_s = \pi a c r_s^2 T_e^4,$$

but this is much less reliable than the other two. If stars really did radiate like black bodies the method would be straightforward. The distribution of radiation with wavelength would give the surface temperature of the star T_*. The amount of radiation received on unit area of the Earth's surface in a given wavelength range could be measured and the radiation by unit area of a body at temperature T_* in the same wavelength range would be determined from the Planck function. The ratio of these two quantities would be r_*^2/d_*^2, where r_* is the radius of the star and d_* its distance. The radius could then be found if d_* were known.

In practice stars do not behave like black bodies. Attempts are made to measure the total radiation from the star falling on unit area of the Earth' surface by means of a bolometer. This is :

$$L_*/4\pi d_*^2 = acr_*^2 T_{e*}^4/4d_*^2. \tag{2.21}$$

If the distance of the star is known, a value is then known for the product $r_*^2 T_{e*}^4$. If the star's radiation is not too different from that of a black body, it may be possible to estimate T_{e*} and hence r_*. However, as the uncertainty in r_* is the uncertainty in T_{e*}^2, this is only a very approximate method of estimating stellar radii.

It should finally be remarked that the Sun is the only star which appears as a disk rather than as a point of light and for the Sun a much more direct measurement of angular diameter and hence radius is possible.

Chemical composition; spectra

In the middle of the last century it was realized that the chemical elements possessed their own characteristic spectra. If heated to incandescence an element would emit radiation of various well-defined frequencies; if the element was placed between the observer and a source of white light, it would absorb light of the same frequencies. Early in this century Bohr produced his model of the atom in which the electrons could exist in various orbits of definite energy around the positively charged nucleus and where energy was either emitted or absorbed when an electron moved from one energy level to another. This gave a natural explanation of the spectral lines of elements. Although later developments of the quantum theory have shown that Bohr's theory is not correct in detail, it is adequately true for our present purposes.

The characteristic spectral lines of many elements and sometimes molecules can be observed in the light received from stars. Sometimes they appear as *emission lines* where the light of a particular frequency is enhanced, but more often they appear as *absorption lines* where the emission from the star at the given frequency is less than that for neighbouring frequencies. Whether they are emission lines or absorption lines, their presence indicates that the element concerned is present in the outer layers of the star. As the distance radiation can travel inside a star before it is absorbed is very small compared to the radius of the star, direct information is only obtained about the chemical composition of the outermost layers from which radiation escapes from the star. As we shall see later, it is believed that the composition of the outer layers can sometimes be very unrepresentative of the star as a whole, although that is usually only the case when significant nuclear reactions have occurred in the star. For most stars the surface chemical composition is similar to that of the star at birth. Soon after the birth of the science of spectroscopy, it was found that most of the chemical elements were present in the outer layers of the Sun. In fact the existence of the element helium was first suggested by spectral lines from the Sun before it had been discovered on Earth.

Spectral types

When the spectra of a reasonable number of stars had been studied, it was found that the stars could conveniently be divided into a number of classes or *spectral types*. The division between the classes was not sharp but for most stars it was reasonably unambiguous. The spectral classes were based on which element was most prominent in the spectra of the stars and these prominent elements varied considerably from star to star. In the Harvard classification the spectral types were denoted by capital letters A, B, C. It was subsequently realized that some of the groups were superfluous and that a more meaningful order for those classes that remained was OBAFGKMRNS.† The main characteristics of these spectral types are shown in Table 2.

Originally it was thought that these observations were closely related to the chemical compositions of the stars and that the most prominent elements in the spectra were the most abundant elements in the stars. Later it was realised that the surface temperature of the stars also played a vitally important role and the order OBA . . . is essentially an order of decreasing surface temperature.

The reason why temperature is very important in determining stellar spectra is as follows. If a particular spectral line is to be absorbed or emitted in a stellar atmosphere, there must be present atoms with electrons in the correct energy levels for the absorption or emission to occur. At low temperatures all of the atoms are in what are known as their *ground states*, with the electrons near to the

Table 2. *Main features in the spectrum of different spectral types.*

O:	Ionised helium and metals, weak hydrogen
B:	Neutral helium, ionised metals, hydrogen stronger
A:	Balmer lines of hydrogen dominate, singly ionised metals
F:	Hydrogen weaker, neutral and singly ionised metals
G:	Singly ionised calcium most prominent, hydrogen weaker, neutral metals
K:	Neutral metals, molecular bands appearing
M:	Titanium oxide dominant, neutral metals
R,N:	CN, CH, neutral metals.
S:	Zirconium oxide, neutral metals.

† The usual way of remembering the present order is through the mnemonic 'Oh be a fine girl kiss me right now, sweetheart'. If this is thought sexist, guy can replace girl.

Table 3. *Effective temperature as a function of spectral type for main sequence stars*

Spectral type	O	B0	A0	F0	G0
T_e/K		50 000	25 000	11 000	7600	6000

Spectral type	K0	M0	M5	R,N,S
T_e/K		5 100	3 600	3 000	3 000

Each spectral type labelled by a capital letter is sub-divided into subclasses labelled by numbers, as in M5 above. The Sun has type G2.

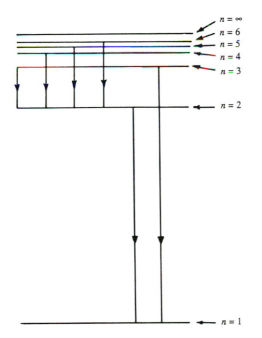

Fig. 20.   The energy levels of the hydrogen atom. The vertical coordinate represents the energy difference between excited states and the ground ($n = 1$) state. The arrowed transitions to the first excited ($n = 2$) state represent emission of the Balmer series.

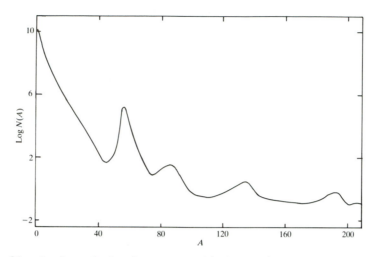

Fig. 21.   A schematic abundance curve. A is the atomic mass number and $N(A)$ is the number of atoms with mass number A; the actual numbers are chosen so that there are 10^6 silicon atoms. The true abundance curve is much more irregular.

nucleus. As the temperature rises, some electrons move in the higher *excited states* and later the atoms are ionised. Hydrogen has only one electron. The spectral lines of hydrogen which fall in the visible region of the spectrum are those which involve transitions down to and up from the first excited state (fig. 20). These lines are known as the Balmer series. These Balmer lines are only strong in stars with intermediate surface temperatures; at low temperatures all of the hydrogen is in the ground state and the only possible absorption lines are in the ultraviolet, while at high temperatures it is mainly ionised. As soon as it was realized that the spectral sequence was primarily a temperature sequence, it became clear that the variations of chemical composition from star to star were mainly slight. The approximate relation of surface temperature to spectral type is shown in Table 3. In the coolest stars of spectral types M–S relatively small abundance differences do have very important effects in the observed spectra as they determine exactly which molecules are formed.

Element abundances

Although the recognition that temperature plays a key role in the appearance of the spectra immediately demonstrated that there was relatively little difference in the chemical composition of stars, there is still considerable difficulty in transforming the raw data of the observations into reliable chemical compositions. There are many factors involved, such as the detailed structure of the star's atomosphere and many atomic properties. As a result deduced chemical compositions are liable to be revised as the theory of the structure of stellar atmospheres is improved. Even in the case of the Sun, which can be observed much more thoroughly than any other star, the estimated abundances of the more abundant elements are still subject to revision. However, the general character of the observations is reasonably clear. Hydrogen is the most abundant element, with only helium at all comparable. There is a gradual decline in the abundances to higher atomic number with a decided local peak in the neighbourhood of iron and with several subsidiary peaks at higher atomic number. A schematic abundance curve is shown in fig. 21 which shows logarithm of abundance as a function of atomic mass number.

The origin of the elements

Although the abundance differences between stars are relatively slight, what differences there are, are very interesting and important. They lead to study of the problem of the origin of the chemical elements. It appears, for example, that the abundances of elements heavier than hydrogen and helium are lower in stars that were formed early in the life of the Galaxy than in stars which have been formed in the recent past. Indeed, it is possible that when the Galaxy was formed it contained no heavy elements and that they have been produced by nuclear reactions in stars in the lifetime of the Galaxy.

Later in the book we shall learn that we expect a succession of nuclear reactions to occur in stars, gradually building light elements into heavier elements. We shall also learn that the more massive a star is, the more rapidly it passes through its life history. Thus the earliest massive stars formed in the lifetime of the Galaxy could have produced the heavy elements which we find in the stars which have been formed more recently, provided that the heavy elements once formed were, at least in part, expelled into interstellar space so that they could be used in the formation of new stars. The problem of deciding whether a very simple initial chemical composition could have been changed into the composition shown in fig. 21, by nuclear reactions in stars in the lifetime of the Galaxy, is a very difficult one. It is outside the scope of the present book although most the subject matter of this book is relevant to it. It is discussed briefly in *Galaxies: structure and evolution*.

Other information from spectral lines

The study of spectral lines can also give much more useful information about the properties of stars. Spectral lines are broadened because the atoms absorbing the radiation do not all have the same speed relative to the observer. This leads to Doppler broadening. The simplest cause is the thermal motions of the atoms but an additional broadening can be produced by convective motions to be discussed in the next chapter. When a star is rotating, one side moves towards the observer and the other side away and rotational Doppler broadening results. A very broad spectral line can also be evidence that a star is losing mass as I shall discuss in Chapter 7. A magnetic field produces a splitting of a spectral line into two or more components through what is called the Zeeman effect. There are some stars which possess very strong magnetic fields, which have a significant influence on their surface properties, but not much influence on the overall structure and evolution of the stars. The magnetic field plays a key role in surface features of the Sun such as sunspots and solar flares. I shall have little to say about stellar magnetic fields in what follows.

General character of the observations

In the earlier part of this chapter I have discussed how the properties of stars can be observed, but I have not discussed the numerical values of mass, radius, luminosity etc. It is often convenient to express the properties of other stars in terms of those of the Sun. These are, with the suffix $\odot$ used to denote the Sun :

$$\left.\begin{aligned} M_\odot &= 1.99 \times 10^{30} \text{ kg,} \\ L_\odot &= 3.86 \times 10^{26} \text{ W,} \\ r_\odot &= 6.96 \times 10^{8} \text{ m,} \\ T_{e\odot} &= 5780\text{K.} \end{aligned}\right\} \tag{2.22}$$

I can now state, in terms of solar values, what ranges of values for M_s, L_s, r_s and T_e have been found in other stars. These are approximately

$$\left.\begin{aligned} 10^{-1} M_\odot &< M_s < 50 M_\odot, \\ 10^{-4} L_\odot &< L_s < 10^{6} L_\odot, \\ 10^{-2} r_\odot &< r_s < 10^{3} r_\odot, \\ 2 \times 10^{3}\text{K} &< T_e < 10^{5}\text{K.} \end{aligned}\right\} \tag{2.23}$$

The very high luminosities of exploding supernovae have been excluded from the above limits as have the properties of the very dense low luminosity neutron stars. It can be seen that there is quite a wide range in the values of all of these quantities, but that the luminosity range is definitely the most extreme. It should be stressed that these numbers refer to stars that have been observed and it is very likely indeed that stars exist with masses, radii and luminosities smaller than those shown in inequalities, (2.23). There may also be a relatively small number of stars which have much larger masses, radii and luminosities.

The Hertzsprung–Russell diagram

Although inequalities (2.23) give an idea of the range of stellar properties, more significant information is obtained by considering the correlation of these properties one with another. One such correlation is shown in a diagram known as the Hertzsprung–Russell diagram. Originally this diagram was a plot of absolute stellar magnitude in some wavelength range against spectral type. It was subsequently realised that variations in spectral type were equivalent to variations in surface temperatures and that the logarithm of surface temperature could replace spectral type. However, as I have stated above, it is difficult to define

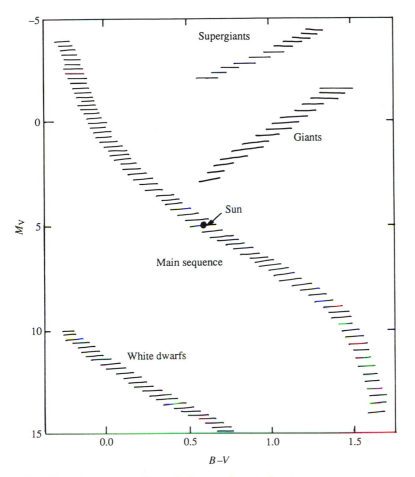

Fig. 22.   The Hertzsprung–Russell diagram for nearby stars.

surface temperature unambiguously and today the observer plots magnitude against colour index, say M_V against $B–V$, where M_V is an absolute magnitude corresponding to the apparent magnitude V. The resulting diagram is also known as a Colour–Magnitude diagram.

When the plot is made for the nearby stars of known distance, a diagram is obtained which is schematically as shown in fig. 22. The stars fall in four main regions. A band which contains the vast majority of the stars is called the main sequence and the other groups are called giants, supergiants and white dwarfs. The latter names arise because giants and dwarfs are found to have large and small radii respectively when these are known. This result also follows because they have larger or smaller luminosities than main sequence stars of the same surface temperature. It is a very important result that the stars do not lie uniformly over the whole of the HR diagram.† The fact that the stars are concentrated in

† In what follows, I shall use the abbreviation HR diagram for Hertzsprung–Russell diagram.

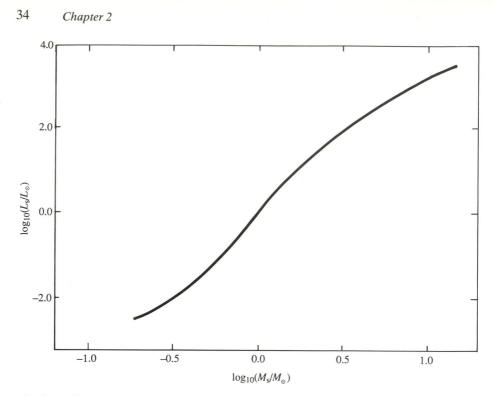

Fig. 23.   The mass–luminosity relation for main-sequence stars.

particular regions with some correlation between magnitude and colour gives some hope that we shall be able to explain the observed properties of individual stars.

As stars must change with time, since they are radiating energy into space, I can ask whether evolution makes any significant difference to their positions in the HR diagram. Clearly at the very beginning and end of a star's life history we may expect its properties to be very different from its properties throughout most of its life, but do these properties change significantly during the major part of its life? In particular, about 90% of nearby stars are main sequence stars. Are *they* main sequence stars for the whole of their active life or are *all stars* main sequence stars for most of their life? It is with questions of this type that I shall be concerned in the later chapters of this book.

The mass–luminosity relation

If I now consider just those main sequence stars for which masses are known, it is found that there is a relationship between mass and luminosity which is illustrated in fig. 23. The more massive stars are the more luminous and the luminosity increases as a reasonably high power of the mass, $L_s \propto M_s^5$, in the steepest part of the curve with an average relation of $L_s \propto M_s^4$. This is another relation which we must hope to understand theoretically. White dwarf stars, for which masses are known, do not obey the main sequence mass–luminosity relation. Although they are very much less luminous than main sequence stars, they have normal stellar masses. As we shall see Chapter 6, there are good theoretical reasons for believing that red giants and supergiants, too, should not obey the main sequence mass–luminosity relation, but at present there is not even one really reliable giant mass known.

Cluster HR diagram

As mentioned earlier, there are only a limited number of nearby stars for which direct measures of distance and hence of absolute luminosity can be made. If we had to rely on these observations, it is unlikely that we should be able to obtain a good theoretical understanding of stellar evolution. What is very useful is the existence of the star clusters which have been mentioned earlier : these include the large globular clusters with perhaps 100 000 or even 1 000 000 stars and the smaller galactic clusters. Both types of cluster are quite compact and physically bound together.

We do not have a direct estimate of cluster distances from the Earth. We can, however, obtain a very crude estimate of their distances by assuming that the brightest cluster stars are generally similar to bright stars in the solar neighbourhood. As we shall soon see, we can improve on this first crude estimate of a cluster distance. The angular diameter of a cluster in the sky gives us a value for the ratio of the diameter of the cluster to its distance from us. For almost all clusters the angular diameter is small, which means that the cluster dimensions are very small compared to their distance from us. This means that all stars in any one cluster are essentially the same distance from us and in addition they probably suffer similar obscuration due to interstellar matter between us and the cluster.

Because the stars in a cluster appear to be physically associated, it is plausible that they were born close together at about the same time. If they were born out of the same cloud of interstellar gas, they may all have essentially the same chemical composition. Thus in trying to understand the properties of stars in a cluster I start by assuming that *all members of a cluster have the same age and chemical composition*. If this is so, the only reason why all of the stars in a cluster are not essentially identical is that they contain different quantities of matter. Thus *the principal factor differentiating stars in a cluster is mass*. This is not to say that other factors do not vary from star to star, but the hope is that these differences are relatively unimportant.

For a cluster of stars an HR diagram can be drawn in terms of apparent magnitude instead of absolute magnitude. When this is done it is found that cluster HR diagrams contain main sequences and giant branches, but that the spread in the diagrams is less than that in the HR diagram for nearby stars (fig. 22). In addition, in many cases there is a continuous transition between the main sequence and the giant branch with very few, if any, stars on the main sequence above the point where the giant branch joins it. The fact that the cluster HR diagrams are rather well defined adds to the hope that the cluster stars do form

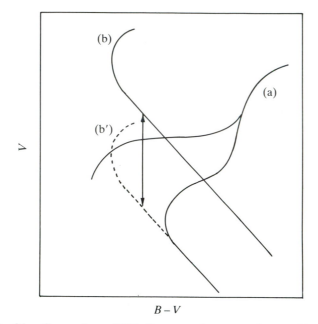

Fig. 24.   Comparison of HR diagrams of two star clusters. Plotted in terms of apparent magnitude V, the main sequences of the two clusters fall in different positions. If it is assumed that they have the same absolute magnitude, the diagram of cluster (b) can be displaced vertically to (b') where its main sequence coincides with that of cluster (a). This main sequence can then be assumed to have the same absolute magnitude as the main sequence in fig. 22.

very homogeneous groups with only mass varying significantly from star to star in any cluster.

It is obviously desirable that we should be able to convert the apparent luminosities of these cluster stars into absolute luminosities, as in any theoretical discussion it is the absolute luminosities which are predicted. Although distances to cluster stars cannot be measured directly, except perhaps for the nearest galactic clusters, there are fortunately ways in which the conversion from apparent to absolute magnitude can be made. Most of the methods rely on some interplay between theory and observation. The simplest approach involves the assumption that the stars of any given colour on a cluster main sequence have the same absolute magnitude as main sequence stars of the same colour in the solar neighbourhood. Because magnitude is related to the logarithm of luminosity, I can convert a cluster HR diagram to absolute magnitude by moving it vertically until its main sequence coincides with the main sequence of the nearby stars; the HR diagram is moved bodily and is not altered in shape in the conversion. This is illustrated in fig. 24. As the width of a cluster main sequence is usually less than the width of the nearby stars main sequence, I could start by making an agreement between the cluster main sequence and the mean line of the sequence for nearby stars. In Chapters 5 and 6 we shall see how this procedure might be improved upon

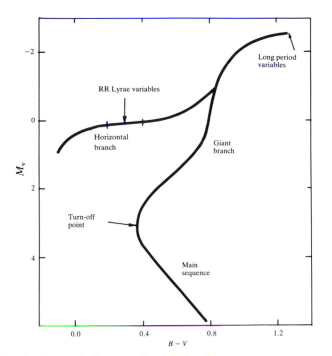

Fig. 25. The HR diagram of a globular cluster.

when we have some idea of what it is that causes a main sequence to have a finite width.

When the conversion from apparent to absolute magnitude has been made, the HR diagrams of globular and galactic clusters appear as in figs. 25 and 26. In fig. 25 is shown the diagram of one typical globular cluster and all globular cluster diagrams are reasonably similar to this one. There is a much greater variety in the HR diagrams of galactic clusters and four of these are sketched in fig. 26. An essential feature of all these diagrams, which has been mentioned above, is that there is a turn-off point from the main sequence. Below this point the cluster has a well-defined main sequence while above the point there are few stars on the main sequence.

Above I have been discussing the *stars in a cluster*, but it should be stressed that it is not really possible to label stars unambiguously as cluster stars. In the direction of a cluster, there are likely to be stars which lie between us and the cluster, and in the case of a nearby cluster there may be many stars visible which lie beyond the cluster. If we are only interested in doing statistics on the number of stars in a cluster without identifying individual cluster members, this can be done by counting stars in a region of the sky near to the cluster and by subtracting a similar number of stars per unit area from the count of cluster stars. For some nearby clusters radial velocities and proper motions can be used to eliminate spurious members. In general more indirect methods must be used and there will always be some uncertainty.

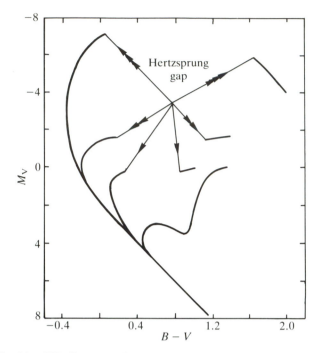

Fig. 26.   HR diagrams of several galactic clusters.

The main features of the cluster HR diagrams shown in figs. 25 and 26 are as follows

Globular clusters

1. They all have a main sequence turn-off in a similar position and a giant branch joining the main sequence at that point.
2. They have a horizontal branch running from near the top of the giant branch to the main sequence above the turn-off point.
3. In many clusters there is a region of the horizontal branch which is populated only by stars of variable luminosity. These are known as RR Lyrae stars after the first star of their type to be studied and they will be discussed on page 42.

Galactic clusters

1. There is considerable variation in the position of the main sequence turn-off point, with the lowest being in about the same position as those of the globular clusters.
2. In many clusters there is a gap between the main sequence and the giant branch, know as the Hertzsprung gap.

Later in this book we shall see how the theory of stellar structure and evolution has

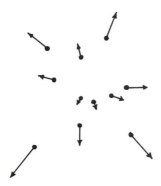

Fig. 27.   The motion of stars in an expanding stellar association; the lengths of the arrows are proportional to the velocities of the stars.

given a general understanding of why globular and galactic cluster HR diagrams have the shape that they have.

Expanding stellar associations

As well as galactic and globular clusters, there are other groups of stars known as expanding associations. These are commonly groups of main sequence O and B stars of high luminosity, which are in the same region of the sky. When their properties are studied, it is found that they appear to be expanding from a common centre in such a way that their velocities are roughly proportional to their distance from the centre (fig. 27). In a typical case, if we extrapolate back their present velocities, they would all have been very close together a few million years ago. Expanding associations also occur containing stars which have irregular variations in their luminosity which are known as T Tauri stars and which will be discussed briefly on page 42 below. It is an obvious question to ask, if the stars in an expanding association were close together a few million years ago, where were they before that? It is believed that before that the stars did not exist as such and that in an expanding association we are observing the after-effects of a multiple star birth in the recent past history of the Galaxy. It is a natural question to ask why the stars which were born together are now separating. It is believed that the explosion of one or more supernovae resulting from rapidly evolving massive stars expelled the remaining gas from the association. As a result the mass of the association was reduced to an extent that its gravitational attraction could no longer confine the stars.

Special types of star

If a composite HR diagram is drawn up containing not only the nearby stars, but also the members of clusters whose distance has been obtained by

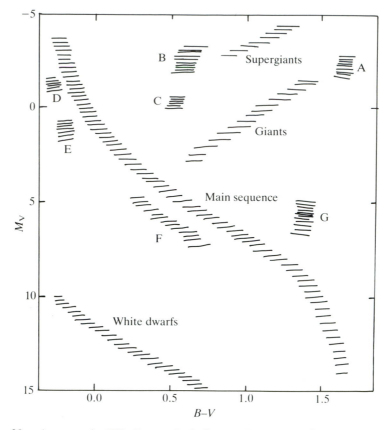

Fig. 28.   A composite HR diagram including, A: long period (Mira) variables, B: cepheid variables, C: RR Lyrae variables, D: Wolf–Rayet stars, E: nuclei of planetary nebulae and old novae, F: subdwarfs and G: T Tauri stars.

indirect means, it looks schematically like fig. 28. In this diagram several types of star are named and I now make a few remarks about each of these. It is first worth mentioning that groups of stars chosen because of characteristics other than luminosity and colour do form fairly compact groups in the HR diagram. I have already mentioned that some stars have variable luminosity and it is found that some regions of the HR diagram contain essentially no non-variable stars while other regions contain no variable stars.

White dwarfs

These are a particularly interesting group of stars about which I shall have more to say in Chapters 8 and 10. The best known white dwarf and the first to be discovered is the binary companion of the apparently brightest star Sirius. Although it is very much less luminous than Sirius, it is of approximately the same colour and it has a mass almost half as great and very similar to that of the Sun.

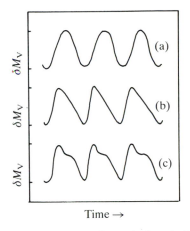

Fig. 29.   Light curves for typical periodic variables. Some like (a) are very symmetrical, many like (b) have a more rapid rise to maximum light than decline to minimum and others like (c) have secondary humps on the light curve.

Unless it has a very peculiar relationship between colour and effective temperature, this means that is must have a very small radius indeed. This follows from (2.7)

$$L_s = \pi a c r_s^2 T_e^4,$$

with the luminosity known and the effective temperature estimated from the colour. All the evidence is that this is true and that *a match-box of white dwarf material would weigh a ton*. If this is true, material in white dwarfs must be at densities orders of magnitude higher than anything we meet or can hope to produce on Earth. This indicates that, if we wish to obtain a theoretical understanding of the structure of stars, we cannot rely on experiments in terrestrial laboratories to give us information about the behaviour of matter in all the conditions that occur in stellar interiors. In many cases we must rely on theory to predict the properties of matter in physical conditions which we cannot check experimentally. In Chapter 10 I shall also discuss neutron stars which are very much denser than white dwarfs. A neutron star of a solar mass has radius of about 10km. Neutron stars are not shown in fig. 28 because they have extremely small luminosities and it is not this normal luminosity which causes them to be discovered or their existence to be deduced.

Variable stars

There are various groups of stars whose light output varies in time and for which it can be shown that the variations are intrinsic to a single star and are not due to an eclipse in a binary system. Some of these stars are very regular and periodic with something near to a sinusoidal variation in light output, others have periodic, but irregular, light curves while others vary much more irregularly. Figure 29 shows the form of some typical light curves for periodic variables. For these the periods vary from a few hours to several hundred days. Several groups of regular variables are shown in fig. 28. These include the RR Lyrae stars with periods of a few hours, the cepheid variables typically with periods of about a week and the long period or Mira variables with periods of up to several hundred days.† The fact that each type of variable is found in a rather compact region in the HR diagram seems very significant. This suggests that light variation is not an accident that can happen to any star, but that a specific combination of physical conditions must be required to make a star vary.

Cepheid variables

The cepheid variables have played a very important role in our understanding of the structure of the Galaxy and the Universe. It was discovered early in the twentieth century that they have a relation between their period and their mean luminosity (the period–luminosity relation) so that, if the period of a cepheid variable is known, so is its absolute magnitude. As the apparent magnitude can be observed an estimate of the distance of the star results. Cepheid variables have been used as *standard candles* in finding the distance to nearby galaxies in which cepheids can be observed. The period–luminosity relation is of course not an exact relation so that there is some uncertainty in the distances obtained. For a long time no cepheid variables were known in star clusters, but now a few are known in galactic clusters. They can be placed in the region of the Hertzsprung gap in the HR diagram. Until the first cepheids were discovered in galactic clusters the period–luminosity relation could not be properly calibrated because there was no cepheid of known distance. Once cepheids were observed in a few clusters of known distance any subsequent discoveries provided an independent distance estimate to their cluster.

The RR Lyrae variables, which have shorter periods than the cepheids, are found in large numbers in globular clusters. All RR Lyrae have a very similar luminosity. This means that the horizontal branches of all globular cluster should occur at about the same luminosity and making them coincide gives a second estimate of the distance of globular clusters. The RR Lyrae stars have lower luminosities than the cepheids, but very similar surface temperature and we shall see later that it is believed that the same physical process causes their variability. The Mira variables with very long periods also occur in globular clusters but I shall not discuss them any further.

As well as regular variable stars there are some stars which vary irregularly in luminosity. Amongst these are the T Tauri variables which occur, as previously mentioned, in expanding associations and are also found just above the lower main sequence in some galactic clusters. They show quite large but irregular variations in luminosity and evidence of outflow of matter from their surfaces. We shall see in Chapter 5 the T Tauri stars are believed to be stars in the process of formation.

† Each of these groups is named after the first star of the type discovered, RR Lyrae, δ Cephei and Mira Ceti.

Novae and supernovae

Other stars which vary very violently and irregularly are the novae and supernovae. They suddenly increase in luminosity by many orders of magnitude. In both cases this increase in brightness is accompanied by an explosive loss of mass from the star. In the case of novae the loss of mass is relatively small and some stars have been novae more than once, but in the case of supernovae the explosion probably shatters the whole star. Only five supernovae have been observed in our Galaxy in the last thousand years, although many others have probably occurred and have been too far away and obscured behind dust clouds to be discovered from the background of faint stars. The explosion of a supernova in the Large Magellanic Cloud, a satellite galaxy of our own Galaxy, in 1987 (SN 1987A) has led to a great increase in observational knowledge of supernovae. Supernovae are so bright that they can be observed in quite distant galaxies, if obscuration due to dust does not intervene. It is difficult to place the stars which become supernovae in the HR diagram because their properties have probably not been studied before they explode. The remnants of supernovae are believed to be neutron stars or black holes which will be discussed in Chapter 10. Possibly in some cases a supernova explosion leaves no remnant behind. Post-novae can be placed in the HR diagram and they are marked in fig. 28. Novae are all members of close binary systems and the nova explosion follows the exchange of mass between the two stars as will be discussed in Chapter 8. Some novae are observed to be recurrent and it is believed that all are and that there is no real distinction between pre- and post-novae.

Planetary nebulae

As well as novae and supernovae, there are also many other stars which are observed to be losing mass at a non-catastrophic rate. In particular there are the planetary nebulae which are shown in fig. 28. These are stars which are surrounded by a sphere or spherical shell of gas which has almost certainly been ejected from the star at a previous stage, because the gas is observed to be expanding away from the star. They are called planetary nebulae because when looked at through a telescope they have a faint greenish disk which looks something like a planet. Stellar structure theory should eventually explain why some stars explode and why others lose mass less violently. Some of the causes of mass loss from stars are discussed in Chapter 7.

Sub-dwarfs

This name is used for all stars which lie significantly below the main sequences defined by the stars in the solar neighbourhood. Some of these stars have additional peculiarities, but the ordinary sub-dwarf apparently has a smaller abundance of elements other than hydrogen and helium than, say, the Sun. In

Chapter 5 we shall see that this lower heavier element content may account for the position of the sub-dwarfs in the HR diagram.

Wolf–Rayet stars

These are very luminous blue stars which are apparently ejecting matter from their surfaces with velocities up to 10^6 m s^{-1}.

It can be seen that many of these special groups of stars are associated with variability of light output and /or stellar instability.

Stellar populations

In 1944, W. Baade introduced the concept that our Galaxy (and other galaxies) was composed of stars of two populations, *population I* and *population II*, and this concept has been important in all subsequent discussions of galactic structure and evolution and stellar evolution. In studying the nearest large galaxy to our own, the Andromeda galaxy M31, and its two companions, he showed that the composite HR diagrams of the companions and of the central regions of M31 resembled that of a globular cluster (fig. 25). In particular the brightest stars in these systems were red supergiants. In contrast the HR diagram of the outer regions of M31 resembled that of a galactic cluster (fig. 26) with the brightest stars being blue main sequence and supergiant stars.

He called the galactic cluster type stars population I stars and the globular cluster type stars population II stars. He found that the central region and the halo region of our Galaxy were like the central regions of the Andromeda galaxy and were composed of population II stars while the disk was made up of population I stars. In addition he found that, from their position in the Galaxy, many of the special groups of stars discussed above could be classified as population I or II. Thus population I included cepheid variables, T Tauri stars, Wolf–Rayet stars and expanding associations, while the gas and dust of the Galaxy was also found in the region occupied by population I stars. Population II included RR Lyrae and Mira variables, planetary nebulae, sub-dwarfs and novae.

In Baade's original classification it was the position of a star in the Galaxy which primarily determined its population, but since then it has become clear that its place of origin is also important. Thus rapidly moving stars in the solar neighbourhood, which is mainly population I, may have been formed in the halo region of the Galaxy and they may be population II stars. Since Baade's original classification two things have become clear. Firstly, there is no really sharp distinction between two populations but there is a gradual transition between two extremes and secondly the main factors distinguishing the two populations are age and chemical composition. This will become clearer later in the book, but the general result is that population I stars are younger and have a larger abundance of heavy elements than population II stars. The fact that the gas and dust in the Galaxy are found in the regions populated by the population I stars means that it is possible for new stars to be formed in these regions.

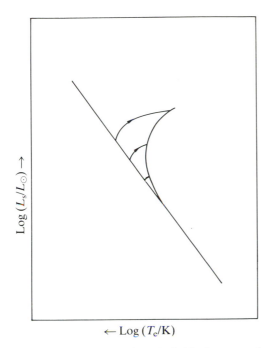

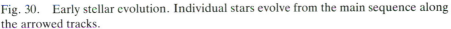

Fig. 30.   Early stellar evolution. Individual stars evolve from the main sequence along the arrowed tracks.

Outline of contents of the following chapters

We shall see in the later chapters that the main task of stellar evolution theory to date has been to try to explain why the HR diagrams of globular and galactic clusters have the particular shape that they have and some of the results of this study will be mentioned here and will be discussed in more detail later. It will be shown in Chapter 3 and 4 that the only physical processes capable of providing energy for stars to radiate for the length of time that they do radiate are nuclear fusion reactions which convert light elements into heavier elements. In Chapter 5 it will appear that the main sequence is populated by stars whose chemical composition is uniform, which are in the stage of burning hydrogen into helium in their interiors. This assumption leads to the prediction of a main sequence and of a main sequence mass–luminosity relation in good qualitative agreement with the observations.

In Chapter 6 it will be shown that, if the conversion of hydrgen into helium leads to the inside of the star becoming rich in helium while the outer regions still retain their initial chemical composition, the star's properties become those of a red giant. Furthermore, it is shown that because the luminosity of a star increases very rapidly with its mass, whilst its supply of nuclear fuel (for given chemical composition) only scales linearly with the mass, the more.massive stars which are higher on the main sequence move into the giant region more rapidly than the less massive stars. This result is shown schematically in fig. 30, which shows how the evolution to the giant branch can lead naturally to the appearance of a turn-off point in the HR diagram. The older a system of stars, the lower the mass of star for which significant evolution has occurred and this suggests that, in observing galactic clusters with very different positions of turn-off point, we are observing systems of very different age.

All of this will be discussed more fully in the chapters which follow but, before I can do this, I must, in the next two chapters, discuss the basic principles governing the structure of a star and what are the important facts of physics which are needed for a discussion of the properties of stars.

Summary of Chapter 2

In this chapter I have discussed the main properties of stars which can, in principle, be deduced from observations. These are mass, radius, luminosity, surface temperature and chemical composition of the outer layers. Some estimate of surface temperature and chemical composition can be made for all stars that are near enough for a detailed study to be made of the distribution with wavelength of the light that they emit. The apparent brightness of a star can always be measured, but this can only be converted into a true luminosity for nearby stars whose distance from the Earth can be measured directly. Masses and radii can only be obtained for a very limited number of stars. Progress in the study of stellar structure and evolution would have been very limited were it not for the existence of regularities in these properties. Thus for most stars there is a definite correlation between values of mass and luminosity and most stars lie in well-defined regions of the Hertzsprung–Russell diagram, which relates luminosity and surface temperature. Real progress in the theoretical interpretation of stellar properties is possible because many stars are members of clusters, which are more homogeneous groups of stars than an arbitary set of nearby stars. Although it is not usually possible to observe all of the properties of individual cluster members, groups of stars, which are similar to nearby stars, can be found in clusters and this enables the distance to a cluster and the luminosity of its stars to be estimated. Throughout the remainder of this book I shall frequently use theoretical results to try to understand the HR diagrams of star clusters.

Also in this chapter I have identified some special groups of stars which have some particular property in common. These include several types of variable stars, novae and supernovae and white dwarfs. As well as giving an understanding of the properties of *ordinary* stars, such as the Sun, theory must also explain why some stars have more unusual properties.

3

The equations of stellar structure

Introduction

In this chapter I consider what are the main physical processes which determine the structure of stars and what equations must be solved in order to find the details of this structure. At the outset it must be stressed that the theoretical astrophysicist does not usually attempt to calculate the properties of a particular star which has been observed. As we have learnt in the last chapter, the number of stars for which there is sufficiently detailed observational knowledge to make this procedure worthwhile is very small. Instead the theoretician tries to isolate the factors which mainly determine the properties of stars and then tries to calculate the structure of a wide range of possible stars. We shall see that the most important factors are the mass and initial chemical composition of the star and the time that has passed since it was formed. In what follows I shall often refer to the *birth of a star*, its *age* and its *chemical composition at birth*. Once calculations have been made for a range of values of mass, chemical composition and age, the results can be compared with the general properties of stars rather than with the properties of individual stars. I shall consider this comparison in Chapters 5 to 10. For one star, the Sun, we possess extremely detailed observational information and there has been a considerable effort to try to obtain a theoretical understanding of its properties.

Observations of the light from stars give information only about the surface layers, and the main object of the theory of stellar structure is to calculate the surface properties of stars with a wide range of mass, chemical composition and age. Although only the surface properties such as radius and surface temperature can be directly compared with observation, I shall show that it is impossible to calculate these without solving equations which also determine the complex internal structure of the star. Thus theory predicts what conditions are like in the deep interiors of stars from which no light can reach us directly: It has, however, been realized that neutrinos which are emitted in nuclear reactions, which are believed to occur near the centre of the Sun, can reach the Earth in very large numbers, and on page 91 I shall discuss attempts which have been made to detect these neutrinos and to obtain information about physical conditions in the centre of the Sun.

A star is held together by the force of gravitation, the attraction exerted on each part of the star by all other parts. If this force alone were important the star would shrink very rapidly, but this attractive gravitational force is resisted by the pressure of the stellar material in the same way that the kinetic energy of the molecules, or equivalently the pressure of the Earth's atmosphere, prevents the atmosphere from collapsing to the surface of the Earth. These two forces, gravitational attraction and thermal pressure, play the principal role in determining the structure of a star. I shall show shortly that they must be almost in balance if the properties of stars are not to change much more rapidly than they are observed to. In addition to discussing the forces which act inside stars, I must also consider the thermal properties of stars. The surface temperatures of stars are high compared to the temperature of their surroundings and they are continually radiating energy into space. If stars are not to change their thermal properties more rapidly than is observed, energy must be continually supplied to make good this loss. I must discuss the origin of this energy and the manner in which it is transported to the surface of the star where it is radiated.

When I discuss the forces acting in stars and the thermal properties of stars, I shall discover that there are three characteristic times entering into the problem. If the gravitational and pressure forces are seriously out of balance, the star contracts or expands significantly in a time, t_d, which I call the dynamical time-scale of the star. The ratio of the total thermal energy of a star to the rate of loss of energy from its surface is called the thermal time-scale, t_{th}. As I shall show later, the ultimate source of the energy radiated by a star is nuclear reactions in the interior. The total nuclear energy resources of a star divided by the rate of energy loss is called the nuclear time-scale t_n. For most stars at most stages in their evolution, the inequalities

$$t_d \ll t_{th} \ll t_n \tag{3.1}$$

are true. These inequalities enable some important approximations to be made in the equations of stellar structure.

In this chapter I make two fundamental assumptions about the structure of stars. I suppose that, although stars do evolve, their properties change so slowly that at any time it is a good approximation to neglect the rate of change with time of these properties. I also suppose that stars are spherical and that they are symmetrical about their centres. If these two assumptions are made, the structure of a star is governed by a set of equations in which all of the physical quantities depend on the distance from the centre of the star alone. I start by making these assumptions and later consider under what conditions they are justified.

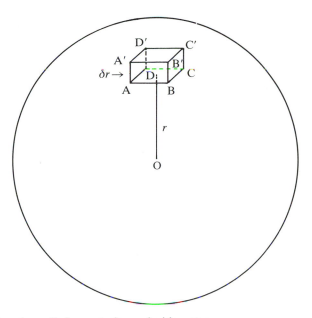

Fig. 31.   A small element of mass inside a star.

Balance between pressure and gravitational forces

Consider first the forces acting on a small element of matter in a spherical star (fig. 31). The two faces ABCD, A′B′C′D′ of the element are both perpendicular to the line joining their centres to the centre of the star O and they are of equal area δS. The lower face is at a distance r from the centre of the star and the upper face is at a distance $r + \delta r$. The volume of the element is $\delta S \delta r$ and provided it is an infinitesimal element its mass differs only slightly from $\rho_r \delta S \delta r$, where ρ_r is the density of stellar material at radius r. The forces acting on the element are the gravitational attraction of the remainder of the star on the element and the forces due to the pressures on the six faces of the element.

The gravitational force acting on an element of mass in a spherical body whose density depends only on distance from the centre takes a particularly simple form. The force acting on the element is the same as if all of the mass interior to the element were concentrated at the centre of the body and all the remainder of the body were neglected. Thus the gravitational force acting on the mass element is $GM_r\rho_r\delta S \delta r/r^2$, where G is the Newtonian gravitational constant and M_r is the mass contained within the sphere of radius r. The gravitational force is directed towards the centre of the star. The expression for the force is accurately true only in the case of an infinitesimal element of matter.

The forces due to the pressures exactly balance apart from the forces on the faces ABCD and A′B′C′D′. The force on ABCD is in the outward radial direction and is $P_r\delta S$ where P_r is the pressure of the stellar material at radius r. In a

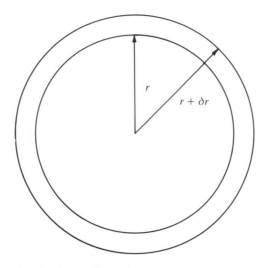

Fig. 32.   A spherical shell inside a star.

similar way the force on A′B′C′D′ is in the inward radial direction and is $P_{r+\delta r}\delta S$, where $P_{r+\delta r}$ is the pressure at radius $r + \delta r$. I can now write down the condition that the net force on the element is zero, i.e. that the star is in equilibrium. Thus

$$P_{r+\delta r}\delta S - P_r\delta S + (GM_r\rho_r/r^2)\delta S\delta r = 0. \tag{3.2}$$

Provided an infinitesimal element is being considered I can write:

$$P_{r+\delta r} - P_r = (\mathrm{d}P_r/\mathrm{d}r)\delta r. \tag{3.3}$$

If (3.3) and (3.2) are combined, I obtain:

$$\frac{\mathrm{d}P_r}{\mathrm{d}r} = -\frac{GM_r\rho_r}{r^2}. \tag{3.4}$$

Equation (3.4) is known as the *equation of hydrostatic support*. In what follows I shall usually omit the suffix r from quantities such as P, M and ρ but it must be remembered that they are all functions of r.

In (3.4) the three quantities M, ρ and r are not independent since the mass contained within a sphere of radius r is determined by the density of the material at points within radius r. A relation between M, ρ and r can be obtained as follows. Consider the mass of a spherical shell between radii r and $r + \delta r$ (fig. 32).

The mass of this shell is approximately $4\pi r^2\rho\delta r$, provided that δr is small. The mass of the shell is also the difference between $M_{r+\delta r}$ and M_r, which for a thin shell can be written:

$$M_{r+\delta r} - M_r = (\mathrm{d}M/\mathrm{d}r)\delta r.$$

Then, equating the two expressions for the mass of the spherical shell, I obtain:

$$\frac{\mathrm{d}M}{\mathrm{d}r} = 4\pi r^2\rho. \tag{3.5}$$

Equation (3.5) may be written alternatively as:

$$M_r = \int_0^r 4\pi r'^2 \rho_{r'} \, dr'. \tag{3.6}$$

I have obtained two of the equations of stellar structure. These are two differential equations for the three quantities P, M and ρ in terms of r. It is clear that I require a further relation between them if I can hope to determine them all. There is one fairly obvious type of relation, the equation of state of the stellar material, analogous to the equation of state of an ideal gas. This will relate the pressure and density but it will also in general introduce another quantity, the temperature T. I shall therefore still require at least one more equation. Before discussing these additional equations, it is possible to obtain some useful general information about the structure of stars on the basis of (3.4) and (3.5) alone. First, however, I will discuss when the two basic assumptions of this chapter are likely to be valid.

Accuracy of the hydrostatic assumption

In the derivation of (3.4) it has been assumed that the forces acting on any element of material in a star are exactly in balance. As we shall see later, during its life history a star undergoes periods of radial expansion and contraction and at these times (3.4) is not accurately true. In these circumstances I can generalize (3.4) as follows. The net force acting on an element must be equated to the product of its mass and acceleration. If a is defined to be the acceleration in the *inward* radial direction a term $\rho_r a \delta S \delta r$ must be introduced on the right-hand side of (3.2) and (3.4) becomes:

$$\rho a = \frac{GM\rho}{r^2} + \frac{\partial P}{\partial r}, \tag{3.7}$$

where the partial derivative, $\partial P/\partial r$, is used as P is now a function of both r and t.

I can now estimate what would happen if the two terms on the right-hand side of (3.7) were slightly out of balance. Suppose that their sum is a fraction λ of the gravitational term so that the inward radial acceleration is a fraction λ of the acceleration due to gravity ($g \equiv GM/r^2$). If the element starts from rest with this acceleration, its inward displacement (s) will be given by:

$$s = \tfrac{1}{2}\lambda g t^2. \tag{3.8}$$

The radius will decrease by 10%, for example, in the time

$$t = \sqrt{\left(\frac{r}{5\lambda g}\right)}. \tag{3.9}$$

At the surface of the Sun $r \equiv 7 \times 10^8$m and $g \equiv 2.5 \times 10^2$ms^{-2} so that

$$t \simeq 10^3/\lambda^{1/2}\text{s}. \tag{3.10}$$

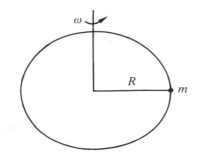

Fig. 33.   A star flattened by rotation.

Since geological evidence concerning the ages of the radioactive elements in the Earth's crust and of fossils suggests that the properties of the Sun have not changed significantly for at least 10^9 years (3×10^{16}s), we can see that at present λ can be no greater than 10^{-27} so that (3.4) must be true to a very high degree of accuracy indeed. To put this another way, if the force of gravitation were not resisted by the pressure gradient of the solar material so that $\lambda = 1$, the radius of the Sun would change significantly in an hour. In the previous chapter I have mentioned that stars do exist in which significant changes occur in hours or days. These include novae, supernovae and some types of variable star. For such stars (3.4) must be replaced by (3.7).

From (3.8) I can obtain an expression for what I earlier called the dynamical time-scale of a star. If I put $s = r$ and $\lambda = 1$, I obtain an estimate of how long it would take the star to collapse completely if pressure forces were negligible. This I define to be the dynamical time, t_d, and is given by:

$$t_d = (2r^3/GM)^{1/2}. \tag{3.11}$$

Validity of assumptions of spherical symmetry

One reason why stars are not accurately spherically symmetrical is that they rotate. Rotating bodies, which are composed of liquids or gases, are flattened at the poles. There is a difference between the polar and equatorial diameter of the Earth which presumably dates from the time when the Earth was molten. In the case of most stars this effect is very small but there are some stars which rotate very rapidly; these are seriously non-spherical and I shall not discuss them in detail in this book. I shall, however, discuss one class of non-spherical star, the close binary star, in Chapter 8.

I can now make an approximate estimate of the importance of rotation in determining the shape of the star. Consider an element of material of mass m near the surface of the star at the equator (fig. 33). In addition to the gravitational and pressure forces, the element will be acted on by an outward force $m\omega^2 R$, where ω is the angular velocity of the star and R is the equatorial radius. This force will be of negligible importance compared to the gravitational force and will therefore not cause a serious departure from spherical symmetry, provided that

$$m\omega^2 R/(GMm/R^2) \ll 1$$

or

$$\omega^2 \ll GM/R^3. \tag{3.12}$$

This expression is closely related to (3.11) and I can say that rotation will have only a slight influence on the structure of a star, provided that the rotation period is very large compared to the dynamical time, t_d.

In the case of the Sun, the effects of rotation are very small. The Sun rotates once in about a month so that $\omega \simeq 2.5 \times 10^{-6} \, \mathrm{s}^{-1}$ and $\omega^2 r^3/GM \simeq 2 \times 10^{-16}$, which suggests that we can reasonably neglect departures from spherical symmetry due to rotation. The rotation of the Sun is interesting and not fully understood. It does not rotate as a solid body and at its visible surface the equatorial regions rotate more rapidly than the polar regions. There was at one time a claim that the Sun was considerably more flattened than could be caused by its surface rotation. This could be the case if the deep interior were rotating much more rapidly than the outside. In turn the gravitational field of the distorted Sun would have influenced the motion of Mercury in such a way that this could no longer be explained by the General Theory of Relativity. Further observations have not confirmed this excess flattening of the Sun.

I now consider stars in which departures from equilibrium and spherical symmetry are unimportant and discuss some consequences of (3.4) and (3.5).

Minimum value for central pressure of a star

Use of (3.4) and (3.5) alone, with no knowledge of the type of material of which a star is composed, enables me to find a minimum value for the central pressure of a star whose mass and radius are known. If I divide (3.4) by (3.5) I obtain:

$$\frac{\mathrm{d}P}{\mathrm{d}r} \bigg/ \frac{\mathrm{d}M}{\mathrm{d}r} \equiv \frac{\mathrm{d}P}{\mathrm{d}M} = -\frac{GM}{4\pi r^4} \tag{3.13}$$

Equation (3.13) can now be integrated with respect to M between the centre of the star and its surface to give:

$$-\int_0^{M_s} \frac{\mathrm{d}P}{\mathrm{d}M} \, \mathrm{d}M = P_c - P_s = \int_0^{M_s} \frac{GM}{4\pi r^4} \, \mathrm{d}M, \tag{3.14}$$

where here, and in what follows, the suffixes c and s refer to the centre and surface of the star. Thus M_s, is the total mass of the star, P_c is the central pressure and P_s the surface pressure.

I can now obtain an underestimate of the integral on the right-hand side of (3.14). At all points inside the star r is less than r_s and hence $1/r^4$ is greater than $1/r_s^4$. This means that

$$\int_0^{M_s} \frac{GM}{4\pi r^4} \, dM > \int_0^{M_s} \frac{GM \, dM}{4\pi r_s^4} = \frac{GM_s^2}{8\pi r_s^4}.$$
(3.15)

Equations (3.14) and (3.15) can now be combined to give:

$$P_c > P_s + GM_s^2/8\pi r_s^4 > GM_s^2/8\pi r_s^4.$$
(3.16)

For the Sun I know accurate values of M_s and r_s and these can be inserted into inequality (3.16) to give:

$$P_{c\odot} > 4.5 \times 10^{13} \, \text{Nm}^{-2}$$
(3.17)

or

$$P_{c\odot} > 4.5 \times 10^8 \, \text{atmospheres}.$$
(3.18)

This is a remarkably powerful result which requires no knowledge of the chemical composition or physical state of the solar material. Clearly, however, it gives some information about the possible physical state of the material at the centre of the Sun. It may seem surprising, in view of the very high value of this pressure, that it is believed that the solar material is gaseous. As we shall see shortly, it is not an ordinary gas.

For stars other than the Sun inequality (3.16) can be rewritten as follows:

$$P_c > (GM_\odot^2/8\pi r_\odot^4)(M_s/M_\odot)^2(r_\odot/r_s)^4.$$

Then, using the solar values in the first expression in brackets:

$$P_c > 4.5 \times 10^{13}(M_s/M_\odot)^2(r_\odot/r_s)^4 \, \text{Nm}^{-2}.$$
(3.19)

The virial theorem

A further consequence of the fundamental equations (3.4) and (3.5) can be found by integrating the equations over the entire volume of the star. From (3.4) and (3.5) I can obtain:

$$4\pi r^3 \, dP = -4\pi r GM\rho \, dr = -(GM/r) \, dM.$$
(3.20)

Integrating (3.20) over the whole star:

$$3\int_{P_c}^{P_s} V \, dP = -\int_0^{M_s} (GM/r) \, dM,$$
(3.21)

where V is the volume contained within radius r. Integrating the left-hand side of (3.17) by parts, the equation can be written:

$$3[PV]_c^s - 3\int_0^{V_s} P \, dV = -\int_0^{M_s} (GM/r) \, dM.$$
(3.22)

The integrated part vanishes at the lower limit of integration because $V_c = 0$. The term on the right-hand side of (3.22) is the negative gravitational potential energy of the star (i.e. apart from the minus sign it is the energy released in forming the star from its component parts dispersed to infinity) and I denote this by the symbol Ω. Noting that $dM = \rho\, dV$, (3.22) can be written:

$$4\pi r_s^3 P_s = 3 \int (P/\rho)\, dM + \Omega. \tag{3.23}$$

If the star were surrounded by a vacuum, its surface pressure would be zero and the left-hand side of (3.23) could be put equal to zero. In fact the surface pressure of a star will not be zero but it will be many orders of magnitude smaller than the central pressure or the mean pressure in the interior. This means that the term on the left-hand side of (3.23) is very small compared to either of the terms on the right-hand side and it can usually be neglected and (3.23) can be written in the approximate form:

$$3 \int (P/\rho)\, dM + \Omega = 0. \tag{3.24}$$

Equation (3.24) is usually known as the *Virial Theorem* and I shall use it frequently later in the book.

The physical state of stellar matter

In the early years of the study of stellar structure there was much discussion about the physical state of matter in stars. It was thought that the stars could not be solid because their temperatures were so high and that they could not be gaseous because their mean densities were too high. It is now believed that they are composed of an almost perfect gas in most circumstances. The perfect gas is, however, unusual in two respects.

The most important respect is that the stellar material is an ionised gas or *plasma*. The temperature inside stars is so high that all but the most tightly bound electrons are separated from the atoms. This makes possible a very much greater compression of the stellar material without deviation from the perfect gas law because a nuclear dimension is 10^{-15}m compared with a typical atomic dimension of 10^{-10}m. The word plasma is the name given to a quantity of ionised gas. It has been recognized in recent years that a plasma can be regarded as a fourth state of matter and that most of the known material in the Universe is in this fourth state. It differs from an ordinary gas because the forces between electrons and ions have a much longer range than the forces between neutral atoms.

The second important difference between most laboratory conditions and conditions in stars is that radiation is in thermodynamic equilibrium with matter in stellar interiors, and its intensity is governed by Planck's law (2.5). Just as the particles in a gas exert a pressure which can be calculated from the kinetic theory of gases by considering collisions of particles with an imaginary surface in the gas, the photons in a Planck distribution exert a pressure known as radiation pressure. At one time it was thought that radiation pressure was of comparable importance to gas pressure in ordinary stars. It is now realised that although there are some exceptional stars in which radiation pressure is of vital importance, it is only of marginal significance in most stars.

From the kinetic theory of gases, the pressure of an ideal classical gas can be shown to have the form:

$$P_{\text{gas}} = nkT, \tag{3.25}$$

where n is the number of particles per cubic metre and k is Boltzmann's constant $(1.38 \times 10^{-23} \text{ JK}^{-1})$. This expression for the pressure can be made to correspond with the usual form for the ideal gas law as follows. If I consider a mass of gas $\mathcal{M}$, of molecular weight m, which occupies a volume v, its pressure is given by:

$$P_{\text{gas}}v = \frac{\mathcal{M}}{m} RT = \frac{\mathcal{M}}{m} N_{\text{A}}kT, \tag{3.26}$$

where R is the gas constant $(8.31 \text{ J mol}^{-1}\text{ K}^{-1})$, N_{A} is Avogadro's number $(6.02 \times 10^{23} \text{ mol}^{-1})$ and $k = R/N_{\text{A}}$. If I consider a cubic metre of gas and note that then $n(\equiv \mathcal{M}N_{\text{A}}/m)$ is the number of particles in a cubic metre, (3.25) results. The corresponding expression for radiation pressure is:

$$P_{\text{rad}} = \tfrac{1}{3}aT^4 \tag{3.27}$$

where a is the radiation density constant $(7.55 \times 10^{-16} \text{ Jm}^{-3}\text{ K}^{-4})$.

Minimum value of mean temperature of a star

In the Sun and many other stars radiation pressure is negligible. I shall now attempt to justify this statement as follows. I shall assume that stars are composed of an ideal classical gas with negligible radiation pressure. I shall then use the Virial Theorem to obtain a *lower bound* to the mean stellar temperature. I already know something about stellar densities from observations of masses and radii. At the temperatures and densities found, it appears that the stellar material would indeed be gaseous and radiation pressure would be negligible.

Consider the two terms in the Virial Theorem:

$$3 \int (P/\rho) \, \text{d}M + \Omega = 0.$$

The magnitude of the gravitational potential energy has a lower bound in terms of the total mass and radius of the star. Thus:

$$-\Omega = \int_0^{M_s} \frac{GM \, \text{d}M}{r}.$$

In this integral r is less than r_s everywhere within the star so that $1/r$ is greater than $1/r_s$. Thus:

$$-\Omega > \int_0^{M_s} \frac{GM \, \text{d}M}{r_s} = \frac{GM_s^2}{2r_s}. \tag{3.28}$$

If the star is assumed to be an ideal gas with negligible radiation pressure, the other term in the Virial Theorem can be written:

$$3 \int_0^{M_s} \frac{P}{\rho} \, dM = 3 \int_0^M \frac{kT}{m} \, dM = \frac{3k}{m} M_s \bar{T}, \tag{3.29}$$

where $\rho = nm$ so that m is now the average mass of the particles in the stellar material and $\bar{T}$ is a mean temperature defined by:

$$M_s T = \int_0^{M_s} T \, dM. \tag{3.30}$$

Combining (3.24) and (3.30) and inequality (3.28)

$$\bar{T} > GM_s m/6kr_s. \tag{3.31}$$

For the Sun I can insert values of the mass and radius into inequality (3.31) and can express the mean particle mass in terms of the mass of the hydrogen atom ($m_H = 1.67 \times 10^{-27}$ kg) to obtain:

$$\bar{T}_\odot > 4 \times 10^6 (m/m_H) \text{K}. \tag{3.32}$$

To obtain a numerical value for this lower limit to the mean temperature, I need a value for m/m_H. As we have learnt in Chapter 2, hydrogen is the most abundant element in stars and for fully ionized hydrogen $m/m_H = \frac{1}{2}$, as there are two particles, one proton and one electron, for each hydrogen atom. For any other element whether fully ionized or not the value of m/m_H is greater; this will be discussed in detail in Chapter 4. Thus I can certainly write:

$$\bar{T}_\odot > 2 \times 10^6 \text{K}. \tag{3.33}$$

This is a very high temperature by terrestrial standards and it is also very much higher than the observed surface temperatures of the Sun and other stars. I also have an estimate of the mean density of the Sun through the relation:

$$\bar{\rho}_\odot = 3M_\odot/4\pi r_\odot^3 \simeq 1.4 \times 10^3 \text{ kgm}^{-3}. \tag{3.34}$$

It is now possible to verify that material with the mean density of the Sun at the mean temperature of the Sun will be a highly ionised gas. The mean density of the Sun is only a little higher than that of water and other ordinary liquids and such liquids turn to gases at temperatures much lower than that given by inequality (3.33). In addition at such a temperature, the average kinetic energy of the particles is higher than the energy required to remove many bound electrons from atoms and the gas will thus be highly ionised. Because the gas is ionised, the distances between particles are much greater than their sizes and corrections to the ideal gas law are small.

I can also estimate the importance of radiation pressure at a typical point in the Sun. Thus from (3.25) and (3.27):

$$\frac{P_{\text{rad}}}{P_{\text{gas}}} = \frac{aT^3}{3nk}. \tag{3.35}$$

With $T \simeq \bar{T} \simeq 2 \times 10^6 \mathrm{K}$ and $n \simeq 2\bar{\rho}/m_{\mathrm{H}} \simeq 2 \times 10^{30}$,

$$\frac{P_{\mathrm{rad}}}{P_{\mathrm{gas}}} \simeq 10^{-4}. \tag{3.36}$$

In this calculation I have underestimated $\bar{T}$ but it certainly appears that radiation pressure is unimportant at an average point in the Sun. It should be stressed that these last two discussions are specific to the Sun. Although many other stars are composed of a near ideal classical gas with negligible radiation pressure, there are stars in which the gas is so dense that corrections to the ideal classical gas law are very important and others in which radiation pressure is important.

It is possible to use inequality (3.31) and two plausible assumptions about the properties of stellar interiors to improve inequality (3.16) relating to the central pressure of a star. I suppose that neither the density nor the temperature of the star increases outwards. As heat flows down the temperature gradient, in a steady state we must expect the temperature to decrease outwards. I shall show later that, if the density increases outwards, the star is unstable to convection which turns over the stellar material. With these two assumptions, I write:

$$P_{\mathrm{c}} = n_{\mathrm{c}} k T_{\mathrm{c}} = k\rho_{\mathrm{c}} T_{\mathrm{c}}/m$$
$$= k(\rho_{\mathrm{c}}/\bar{\rho})(T_{\mathrm{c}}/\bar{T})\bar{\rho}\bar{T}/m. \tag{3.37}$$

Now

$$\bar{\rho} = 3M_{\mathrm{s}}/4\pi r_{\mathrm{s}}^3 \tag{3.38}$$

and using this and inequality (3.31) for $\bar{T}$, (3.37) becomes

$$P_{\mathrm{c}} > (GM_{\mathrm{s}}^2/8\pi r_{\mathrm{s}}^4)(\rho_{\mathrm{c}}/\bar{\rho})(T_{\mathrm{c}}/\bar{T}). \tag{3.39}$$

For most stars $\rho_{\mathrm{c}}/\bar{\rho}$ and $T_{\mathrm{c}}/\bar{T}$ are both significantly greater than one and this estimate of the central pressure is much higher than the previous one.

The source of stellar energy

So far I have really only considered the dynamical properties of a star. However, perhaps the most important property of a star is that it continuously radiates energy into space and I must concern myself with the origin of that energy and how it is transported from its place of origin to the surface of the star. Let me first consider the origin of this energy, where, of course, I do not mean the appearance of energy from nothing, but its conversion from another form in which it is not immediately available for the star to radiate. Once again I take the Sun as an example. The Sun radiates energy at a rate of $4 \times 10^{26} \mathrm{Js}^{-1}$ ($4 \times 10^{26}\mathrm{W}$). Using Einstein's relation between mass and energy, $E = mc^2$, this means that the Sun is losing mass at the rate of $4 \times 10^9 \mathrm{kgs}^{-1}$. By studying the radioactive elements in the Earth's rocks and their decay products, it is possible to estimate how long the rocks have been solid. Study of the fossils in the rocks indicates how long living things have been present on Earth. These studies show that the Sun's luminosity cannot have changed significantly in the last few thousand million years and in that time the total mass loss must have been about $2 \times 10^{-4} M_\odot$.

What can have been the source of this energy? Perhaps the simplest idea is that the Sun became very hot at some time in the remote past, perhaps it was created very hot, and has since been cooling down. I can test the plausibility of this by asking for how long the present thermal energy content of the Sun could supply its present rate of energy loss. Another possibility which was seriously considered when the structure of stars was first studied was that the Sun was slowly contracting with a consequent release of gravitational potential energy and that this energy was converted into the radiation which escaped from the surface.

The thermal energy and the gravitational energy of a star composed of an ideal gas are very closely related. In an ideal gas the total thermal energy is obtained by multiplying the number of particles by the number of degrees of freedom, n_f, possessed by each particle and by $kT/2$. Thus the thermal energy per unit volume is $n n_f kT/2$. The number of degrees of freedom n_f is related to the ratio of specific heats γ of the material by $\gamma = (n_f + 2)/n_f$, where γ is the ratio of the specific heat at constant pressure to the specific heat at constant volume. Using the expression (3.25) for the pressure of an ideal gas and introducing the thermal energy per kilogram, u, instead of the thermal energy per unit volume,

$$u = P/(\gamma - 1)\rho. \tag{3.40}$$

The Virial Theorem (3.24) can then be written

$$3(\gamma - 1)U + \Omega = 0 \tag{3.41}$$

for an ideal gas with negligible radiation pressure, where U is the total thermal energy of the star. As mentioned earlier, the material inside a star is highly ionised. A fully ionised gas is a monatomic gas for which the value of γ is 5/3; for such a value of γ (3.41) can be written

$$2U + \Omega = 0. \tag{3.42}$$

Thus for such a star the negative gravitational energy is just equal to twice the thermal energy.

It is clear from (3.42) that the time for which the present thermal energy of the Sun can supply its radiation and the time for which the past release of gravitational potential energy could have supplied its present rate of radiation differ only by a factor of 2, and to get an approximate idea of the time only one need be considered. The total release of gravitational potential energy is of order $(GM_\odot^2/r_\odot)$J and this would have been sufficient to provide radiant energy at a rate $L_\odot$W for a time:

$$GM_\odot^2/r_\odot L_\odot \simeq 10^{15}\text{s} \simeq 3 \times 10^7 \text{ years.} \tag{3.43}$$

This means that, if the Sun's radiation were supplied by either contraction or cooling (I can consider either since the factor of 2 is insignificant), it would have changed substantially in the last ten million years, while geologists tell us that it can hardly have altered in a time a hundred times longer. The time

$$t_{th} \equiv GM_s^2/r_sL_s, \tag{3.44}$$

for any star, is the thermal time-scale introduced in (3.1).

It is clear that I must look for another source for the Sun's radiant energy, but before doing that I can deduce another very important result from (3.42). The total energy of a star can be defined by:

$$E = U + \Omega, \tag{3.45}$$

provided that there are no other sources of energy. If the star radiates energy into space, E must decrease. Combining (3.42) and (3.45) we have:

$$E = -U = \Omega/2. \tag{3.46}$$

The total energy of the star is negative and it is equal to half the gravitational energy or equivalently minus the thermal energy. We now see that a decrease in E leads to a decrease in Ω but a increase in U. Thus a star composed of a perfect gas, with no hidden energy supplies, contracts and heats up as it radiates energy. We thus have the rather paradoxical result that such a star finds it difficult to cool down; any attempt to lose energy causes the star to contract and to release energy at a rate which not only supplies the energy loss from the surface, but also heats up the material of the star. Although I have obtained this result for a fully ionised gas, it is true provided the ratio of specific heats γ exceeds 4/3. It is a very important result which I shall refer to when I consider the way in which a star evolves. We now see that, if the Sun is an ideal classical gas, it is impossible for cooling to be supplying its surface energy loss, regardless of the geological considerations.

I now return to a consideration of the source of the energy which the Sun and other stars radiate. If it is neither gravitational energy or thermal energy, it seems that it must be released by the conversion of matter from one form to another. Moreover it must be capable of releasing at least 2×10^{-4} of the rest mass energy of the Sun (see page 59). This immediately rules out chemical reactions such as the combustion of coal, gas and oil which only release up to 5×10^{-10} of the rest mass energy. In fact the only known way in which quantities of energy as large as this can be released, by the change of matter from one form to another, is through nuclear reactions. These can be either fission reactions of heavy nuclei such as occur in the atomic bomb and nuclear reactors and which can release 5×10^{-4} of the rest mass energy or fusion reactions of light nuclei which occur in the hydrogen bomb and can release at most almost 1% of the rest mass energy. Not only are fusion reactions capable of a higher energy release, but also, as we have seen in Chapter 2, the light elements are the more abundant. It is now believed that nuclear fusion reactions are the source of the energy radiated during most phases of a star's evolution. This will be discussed more fully in the next chapter.

Although gravitational energy cannot be powering the Sun at present, it will have been important before the centre of the Sun became hot enough for nuclear reactions to occur in its pre-main-sequence evolution. In addition there are very

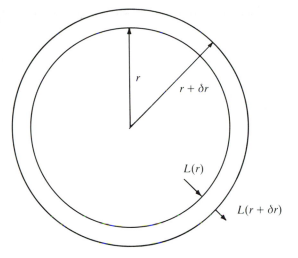

Fig. 34.   Energy release in a spherical shell. $L(r + \delta r)$ exceeds $L(r)$ by the energy released in the shell.

compact stars, *neutron stars*, which will be discussed in Chapters 8 to 10, whose formation involves a release of gravitational energy which is a significant fraction of the rest mass energy. The ratio of the magnitude of the gravitational energy to the rest mass energy is approximately $(GM_s^2/r_s)/M_sc^2$ which becomes significant if the radius approaches GM_s/c^2. A neutron star can have radius less than ten times this value. I shall explain in Chapter 10 that the general theory of relativity provides a critical radius of $2GM_s/c^2$ within which stars become *black holes*.

Relation between energy release and energy transport

Assuming that time variations are unimportant, I can immediately write down one equation relating the rate of energy release and the rate of energy transport. I suppose as before that the star is spherically symmetrical so that all energy is transported in the radial direction. Furthermore, the energy sources are distributed in such a way that the rate of energy release at radius r is $\varepsilon(r)\mathrm{Wkg}^{-1}$. Suppose energy flows across a sphere of radius r at a rate $L_r\mathrm{W}$. Then I can equate the difference between the energy crossing a sphere of radius $r + \delta r$ and a sphere of radius r to the energy released in the spherical shell (fig. 34). Thus:

$$L_{(r+\delta r)} - L_r = 4\pi r^2 \delta r \rho \varepsilon$$

or

$$\frac{\mathrm{d}L}{\mathrm{d}r} = 4\pi r^2 \rho \varepsilon. \tag{3.47}$$

In deriving this equation I have again neglected the change with time of the stellar properties. Thus, for example, I have not allowed for the possibility that some of the energy released in the shell is used to heat up or change the volume of the shell.

This neglect of the time dependence will normally be justified if the energy sources at present being used are capable of supplying the star's radiation for a time long compared with its thermal time (3.44); in the notation of (3.1) this means:

$$t_n \gg t_{th}. \tag{3.48}$$

This is by no means always true and it is much more likely that time dependence has to be included in (3.47) than in (3.4).†

Method of energy transport

I have obtained one further equation for the structure of a star, but only by introducing two more unknown quantities ε and L, so that it is clear that several more equations are still required. I must now consider the way in which energy is transported outwards in a star. There are three possible modes of energy transport; conduction, convection and radiation. Of these, there is no real distinction in principle between conduction and radiation, which both depend on the collision of energetic *particles* with less energetic *particles* resulting in an exchange of energy.

In the case of conduction, energy is mainly carried by electrons. The more energetic electrons from the hotter regions collide with electrons from the cooler regions and in the collision transfer energy. In the case of radiation the energy is carried by light quanta (photons). In most stars gas pressure is more important than radiation pressure (as I have shown for the Sun on page 58) and the same is true of the energy density. Thus:

$$\left. \begin{array}{ll} P_{gas} = nkT, & \rho u_{gas} = \frac{3}{2}nkT, \\ P_{rad} = \frac{1}{3}aT^4, & \rho u_{rad} = aT^4, \end{array} \right\} \tag{3.49}$$

where ρu is the energy per unit volume and the gas has been assumed to be monatomic. As the thermal energy of the electrons is much greater than the energy of the photons, it might be expected that thermal conduction would be the more important mechanism of energy transport of the two. However, there are two factors which determine the efficiency of transport of energy, the energy content, and the distance particles travel between collisions. The latter is known as the *mean free path*. If the mean free path is large, particles can get from a point where the temperature is high to one where it is significantly lower before colliding and transferring energy and a large transport of energy results. Although neither

---

† It is possible to generalize (3.47) to take account of energy release which increases the thermal energy, or does work in changing the volume of an element of stellar material. The generalized equation is:

$$\frac{du}{dt} + P\frac{dv}{dt} = \varepsilon - \frac{1}{4\pi r^2 \rho}\frac{\partial L}{\partial r}.$$

where the symbol d/dt refers to the rate of change with time of the properties of a fixed element of material, u is the thermal energy per kg and v is the specific volume $(1/\rho)$. This equation has been used in the studies of stellar evolution, which will be described in the later chapters of this book.

electrons nor photons can travel very far without collisions in typical conditions in stars, the photons have a considerably longer mean free path than the electrons and this more than offsets the greater total energy possessed by the electrons. In most stars the amount of energy carried by conduction is negligible compared to that carried by radiation. I shall explain in Chapter 10 that conduction is important in the final stage of stellar evolution.

Even so, it is possible to give an argument which suggests that the mean free path of photons in the Sun must be very small and that a photon must suffer many collisions in its journey from the interior to the surface of the Sun. Energy is released by nuclear reactions in the central regions of the Sun. If the photons released in these nuclear reactions travelled with the velocity of light to the surface of the Sun, they would escape from the Sun in a little over 2 s. In fact the energy released at the centre of the Sun slowly diffuses outwards. We have seen previously that the total thermal energy of the Sun could supply its rate of radiation for about 3×10^7 years. This gives us an estimate of how long it takes a photon to diffuse from the centre of the Sun to the surface. When we observe the energy radiated at the solar surface we are usually seeing the results of nuclear reactions which occurred some tens of millions of years ago. Of course this is a rather crude discussion because photons do not retain their identity and what I have called collisions include absorption of radiation. However, it is certainly true that, if the Sun had stopped releasing energy from nuclear reactions about ten million years ago, we should only be starting to notice the consequences now.

Whether conduction or radiation is being considered, the flux of energy per square metre per second (F) can be expressed in terms of the temperature gradient and a coefficient of thermal conductivity (λ). Thus:

$$\left.\begin{aligned} F_{\text{cond}} &= -\lambda_{\text{cond}}\, \mathrm{d}T/\mathrm{d}r, \\ F_{\text{rad}} &= -\lambda_{\text{rad}}\, \mathrm{d}T/\mathrm{d}r, \end{aligned}\right\} \tag{3.50}$$

in spherical symmetry, where the minus sign indicates that heat flows down the temperature gradient and, for example, $L_{\text{rad}} = 4\pi r^2 F_{\text{rad}}$. The thermal conductivity measures the readiness of heat to flow. The astronomer usually works in terms of a quantity which he calls the *opacity* of stellar material which measures the resistance of the material to the flow of heat. The opacity is defined by the relation:

$$\kappa = \frac{4acT^3}{3\rho\lambda}, \tag{3.51}$$

so that

$$F_{\text{rad}} = -\frac{4acT^3}{3\kappa_{\text{rad}}\rho}\frac{\mathrm{d}T}{\mathrm{d}r} \tag{3.52}$$

and

$$F_{\text{cond}} = -\frac{4acT^3}{3\kappa_{\text{cond}}\rho}\frac{\mathrm{d}T}{\mathrm{d}r}. \tag{3.53}$$

Fig. 35.   A layer of liquid heated from below.

With this definition of κ, the probability that a photon will be absorbed in travelling a distance δx is $\kappa \rho \delta x$. The derivation of (3.52) is discussed in a little more detail in Appendix 2.

If all the transport of energy is by radiation and conduction, (3.52) and (3.53) can be combined to give:

$$L = 4\pi r^2 (F_{\text{cond}} + F_{\text{rad}}) = -\frac{16\pi acr^2 T^3}{3\kappa\rho}\frac{\mathrm{d}T}{\mathrm{d}r}$$

or

$$\frac{\mathrm{d}T}{\mathrm{d}r} = -\frac{3\kappa L\rho}{16\pi acr^2 T^3}, \tag{3.54}$$

where

$$\frac{1}{\kappa} = \frac{1}{\kappa_{\text{rad}}} + \frac{1}{\kappa_{\text{cond}}}. \tag{3.55}$$

Equation (3.54) can also be rewritten:

$$\frac{\mathrm{d}P_{\text{rad}}}{\mathrm{d}r} = -\frac{\kappa L\rho}{4\pi cr^2}, \tag{3.56}$$

which is formally rather similar to (3.4):

$$\frac{\mathrm{d}P}{\mathrm{d}r} = -\frac{GM\rho}{r^2}.$$

Clearly the flow of energy by radiation and conduction can only be determined if an expression for κ is available. How such an expression can be obtained will be discussed in Chapter 4

Energy transport by convection

Energy transport by conduction and radiation occurs whenever a temperature gradient is maintained in any body. The same is not true of convection which only occurs in liquids and gases, when the temperature gradient exceeds some critical value. Before I discuss convection in stars I will consider the much simpler problem of convection in a liquid which can be studied in a laboratory. Suppose (fig. 35) that liquid is contained between two parallel surfaces which are maintained at different temperatures T_1 and T_2 with the lower surface being the hotter. For small temperature differences between the two walls, heat is only carried by conduction and the rate of flow of heat is given by the first of the two equations (3.50). If the temperature difference is increased, there is a critical stage when mass motions or convection occur in the liquid and the amount of energy transported increases sharply. At first the motions are fairly regular. The rising and falling elements of fluid form a fairly simple geometrical pattern; in special cases hexagonal convection cells are formed, with fluid rising at the centre of the hexagons and falling at the edges. If the temperature difference is increased still further, the regular patterns disappear and the motions become confused and irregular (turbulent).

Convection starts because the state of the fluid without motions is unstable. Suppose I ask what happens to a small element of fluid which moves upward from the lower surface. When this element has risen a small distance it is hotter than its surroundings and because of its coefficient of thermal expansion it is lighter than its surroundings. As it is lighter than its surroundings it will tend to rise further, but at the same time it will tend to conduct heat to its surroundings and cool down and the frictional force acting on the element will tend to slow it down. If the temperature gradient is high the buoyancy force will *win* and convective motions will start, whereas, if the temperature gradient is lower, the conduction of heat from the element and the resistance to motion offered by the surroundings are sufficient to prevent convection from starting.

Theoretical calculations have predicted that whether or not convective motions will occur should depend on the value of a dimensionless quantity known as the Rayleigh number. This is defined by:

$$R = g\alpha\beta d^4/\lambda\eta, \tag{3.57}$$

where g is the acceleration due to gravity, α is the coeficient of thermal expansion, λ is the coefficient of thermal conductivity, η the kinematic viscosity, d the depth of the layer of liquid and $\beta = |dT/dz|$. It is predicted that convection will occur if

$$R \gtrsim 1700, \tag{3.58}$$

and this has been confirmed experimentally with a variety of liquids. Once the Rayleigh number exceeds this critical value and convection is occurring, I am interested in how the amount of energy carried by convection is related to the Rayleigh number R. Here theory and observation are not in such good agreement. The amount of heat carried by convection when R is large can be measured, but at present there is no theory of fully developed convection which predicts the heat flow completely accurately.

Convection in stars is different from convection in a laboratory in several ways. In the first place there are no rigid walls maintained at well-defined temperatures; in fact, the size of a star and the values of the physical quantities within it depend on the way in which heat is transported. This means that I cannot first calculate the structure of the star and then ask how much energy is carried by convection. In the second place a star is composed of a highly compressible gas rather than an almost

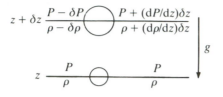

Fig. 36.   A rising convective element.

incompressible liquid. This means that a rising element of fluid in a star has a density which depends not only on its temperature, but also on its pressure, which is also the pressure of its surroundings. In a liquid convection would occur for any temperature gradient at all in a fluid heated below, if it were not for heat flow to the surroundings and the viscosity of the fluid. This is not true in a gas. In this case too, convection will occur if the rising element is lighter than its surroundings but this will depend on two things; the rate at which the element expands due to the decreasing pressure exerted on it and the rate at which the density of the surroundings decreases with height. Thus there must be a finite temperature gradient in a gas before convection will start, whatever the values of its thermal conductivity and viscosity. In the first instance I neglect the effect of heat losses and viscosity.

Condition for occurrence of convection

If a rising element loses no heat to its surroundings (this is called moving adiabatically) its pressure P and volume v obey the relation:

$$Pv^\gamma = \text{const}, \tag{3.59}$$

where, as before γ is the ratio of the two principal specific heats. Equivalently (3.59) can be written:

$$P/\rho^\gamma = \text{const}. \tag{3.60}$$

Suppose an element rises a distance δz (fig. 36), starting with pressure P and density ρ and finishing with pressure $P - \delta P$ and density $\rho - \delta \rho$. Its undisturbed surroundings at the upper level will have pressure, $P + (dP/dz)\delta z$, and density, $\rho + (d\rho/dz)\delta z$, where dP/dz is negative as gravity acts downwards. From (3.60) it can be seen that

$$(P - \delta P)/(\rho - \delta \rho)^\gamma = P/\rho^\gamma. \tag{3.61}$$

In (3.61), if the distance δz and the changes δP, $\delta \rho$ are assumed small, $(\rho - \delta \rho)^\gamma$ can be replaced to sufficient accuracy by $\rho^\gamma - \gamma \rho^{\gamma-1}\delta \rho$ and (3.61) becomes:

$$\delta P = (\gamma P/\rho)\delta \rho. \tag{3.62}$$

As the element rises, it will remain at the same pressure as its surroundings, because pressure variations can be smoothed out at the speed of sound, so that

$$\delta P = (-dP/dz)\delta z. \tag{3.63}$$

Combining (3.62) and (3.63), I obtain:

$$\delta\rho = (\rho/\gamma P)(-dP/dz)\delta z. \tag{3.64}$$

The rising element will then be lighter than its surroundings and will continue to rise if

$$\rho - \delta\rho < \rho + (d\rho/dz)\delta z$$

or

$$(\rho/\gamma P)(dP/dz) < d\rho/dz, \tag{3.65}$$

where I have used (3.64) and have divided the inequality by δz which is positive. I next divide both sides of (3.65) by dP/dz. As dP/dz is negative, I must also change the sign of the inequality to give

$$\frac{P}{\rho}\frac{d\rho}{dP} < \frac{1}{\gamma}. \tag{3.66}$$

For an ideal gas in which radiation pressure is negligible, $P = \rho kT/m$, where m is the mean mass of the particles in the gas. Provided I am not considering a region in which ionisation or dissociation is taking place and where m would vary with position

$$\log P = \log \rho + \log T + \text{const}$$

and this can be differentiated to give

$$\frac{dP}{P} = \frac{d\rho}{\rho} + \frac{dT}{T}. \tag{3.67}$$

Then (3.66) and (3.67) combine to give the condition for occurrence of convection as

$$\frac{P}{T}\frac{dT}{dP} > \frac{\gamma - 1}{\gamma}. \tag{3.68}$$

In deriving this condition I have not considered the spherical geometry of a star. This is not, however, a defect as the criterion for the onset of convection only depends on conditions in the immediate neighbourhood of the element concerned and in a small enough region it is impossible to distinguish between a plane and a spherical system.†

---

† This corresponds to the use of plane geometry on small regions of the Earth's surface.

(3.68) can also be written:

$$\left|\frac{dT}{dz}\right| > \left(\frac{\gamma - 1}{\gamma}\right)\frac{T}{P}\left|\frac{dP}{dz}\right|, \tag{3.69}$$

where the modulus signs are used because both dP/dz and dT/dz are negative, and it can be seen that convection will occur if $|dT/dz|$ exceeds a certain multiple of $|dP/dz|$. As in the case of a liquid a somewhat higher temperature gradient than this will be required before convection occurs because of effects of heat flow from the element and viscosity. However, this correction to the critical temperature gradient is usually small compared to the gradient itself and I shall ignore it in what follows.

Once again I have determined when convection is likely to occur, but I also need to know how much energy will be carried by convection when criterion (3.68) is thoroughly violated. It is not possible to do experiments which mimic the conditions inside stars and at present there is not a generally accepted theory which calculates how much energy will be carried by fully developed convection. Fortunately, as we shall see later, there are occasions when we can manage without this knowledge, but *the lack of a good theory of convection is one of the worst defects in our present studies of stellar structure and evolution.*

The structure of stars

If for a moment I assume that convection does not occur, I have four differential equations governing the structure of stars:

$$\frac{dP}{dr} = -\frac{GM\rho}{r^2}, \tag{3.4}$$

$$\frac{dM}{dr} = 4\pi r^2\rho, \tag{3.5}$$

$$\frac{dL}{dr} = 4\pi r^2\rho\varepsilon \tag{3.47}$$

and

$$\frac{dT}{dr} = -\frac{3\kappa L\rho}{16\pi a c r^2 T^3}. \tag{3.54}$$

Included in these equations are three quantities, P, κ and ε, which I shall consider further in the next chapter. I can say now that, if the star is in a steady state and in a state close to thermodynamic equilibrium, all of these quantities should depend on the density, temperature and chemical composition of the star where, of course, all of these will in general be functions of the radius r. I have already written down in (3.25) and (3.27):

$$P_{gas} = nkT, \qquad P_{rad} = \tfrac{1}{3}aT^4,$$

possible expressions for the pressure and it is a problem of basic physics to tell us what are the pressure, opacity and rate of energy generation of a medium for given conditions of density and temperature. This I shall discuss further in the next chapter, but at present I can assume that such expressions can be obtained and write:

$$P = P(\rho, T, \text{composition}), \tag{3.70}$$

$$\kappa = \kappa(\rho, T, \text{composition}) \tag{3.71}$$

and

$$\varepsilon = \varepsilon(\rho, T, \text{composition}). \tag{3.72}$$

Given the chemical composition of the star I now have seven equations for the seven unknowns P, ρ, T, M, L, κ and ε as functions of r.

Calculations of the structure of a star now involve obtaining expressions for P, κ and ε and then the solution of the four differential equations (3.4), (3.5), (3.47) and (3.54). In general such a solution can only be obtained with the aid of a computer and, if the best possible expressions are used for P, κ and ε, quite a large computer is required. Before the equations can be solved I must consider what are known as boundary conditions. If I consider a single first-order differential equation of the form

$$\frac{dy}{dx} = f(x, y),$$

it will not have a unique solution but will usually have a solution containing one arbitrary constant. The solution can be made precise if I know in advance the value that y must take at one specific value of x. In many physical problems I know the value of y at one of the boundaries of the system and this is then known as a boundary condition. In the problem which I am studying in this chapter, I have not one but four first-order differential equations and I can expect to be able to satisfy four boundary conditions. Indeed, unless I have four boundary conditions, it will be impossible to determine the structure of the star uniquely.

Two of the boundary conditions are at the centre of the star and two at the surface. The two at the centre are quite obvious. The mass and the luminosity are both quantities which must increase outwards from zero at the centre of the star. Thus:

$$M = 0, \qquad L = 0 \quad \text{at} \quad r = 0. \tag{3.73}$$

The surface boundary conditions are not so clear, but simple approximations to these conditions are often used. If stars were truly isolated bodies, we would expect the density and the pressure to fall to zero at the surface. In fact there is no sharp edge to a star, but the density of the Sun near to the visible surface is estimated to be about 10^{-4}kg m^{-3} which is extremely small compared to the value of the mean density $1.4 \times 10^3 \text{ kg m}^{-3}$ given by (3.34). Similarly the surface temperatures of stars are very much smaller than their mean temperatures. In the case of the Sun, its surface temperature of 5780 K can be compared with the typical temperature 2×10^6K predicted by (3.33). Since the surface temperature and density are both very small compared to typical values of these quantities, it seems plausible that the solution of the equations of stellar structure throughout most of the interior of a star will not be seriously affected if the true surface boundary conditions are replaced by the assumption that the density and temperature vanish at the surface. Thus I take:

$$\rho = 0, \qquad T = 0 \quad \text{at} \quad r = r_s. \tag{3.74}$$

Clearly the use of these approximate boundary conditions will not allow me to obtain detailed information about the properties of the outer layers of a star. However, in many cases the radius and luminosity of a star obtained using the correct boundary conditions differ negligibly from those obtained using the approximate boundary conditions. There are also some occasions when the boundary conditions (3.74) give a poor representation not only of the surface layers, but also of the whole internal structure of the star. Such a case will be discussed in Chapter 5. It might be thought that the use of a boundary condition such as (3.74) makes impossible the determination of the surface temperature of a star, which is one of the quantities which our theory should predict. However, once I know the luminosity and radius of a star I can calculate the effective temperature, which is the temperature of a black body with the same radius and luminosity as the star. Thus, from Chapter 2:

$$L_s = \pi a c r_s^2 T_e^4. \tag{2.7}$$

This effective temperature can be compared with observational estimates of surface temperature. The uncertainty involved in converting theoretical effective temperatures into observed colour indices has already been discussed in Chapter 2.

Use of mass as an independent variable

In the introduction to this chapter, it has been explained that the theoretical astrophysicist does not usually try to calculate the properties of a particular star. Instead he studies a wide variety of possible stars. These possible stars can be defined by their *mass* and *chemical composition*: the amount of matter they contain and what kind of matter it is. From the theoretical point of view the mass of a star is to be regarded as something to be chosen before the equations of stellar structure are solved, while the radius is something to be determined from these calculations. For this reason it is often inconvenient that the boundary conditions (3.74) have to be applied at a radius which is not known in advance.

To avoid this difficulty, it is often useful to write the equations in terms of M as the independent variable instead of r. Thus division of (3.4), (3.47) and (3.54) by (3.5) and the inversion of (3.5) itself, give:

$$\frac{dP}{dM} = -\frac{GM}{4\pi r^4}, \tag{3.75}$$

$$\frac{\mathrm{d}r}{\mathrm{d}M} = \frac{1}{4\pi r^2 \rho}, \tag{3.76}$$

$$\frac{\mathrm{d}L}{\mathrm{d}M} = \varepsilon, \tag{3.77}$$

$$\frac{\mathrm{d}T}{\mathrm{d}M} = -\frac{3\kappa L}{64\pi^2 acr^4 T^3}. \tag{3.78}$$

Similarly the boundary conditions (3.73), (3.74) can be written:

$$r = 0, \qquad L = 0 \quad \text{at} \quad M = 0 \tag{3.79}$$

and

$$\rho = 0, \qquad T = 0 \quad \text{at} \quad M = M_\text{s}, \tag{3.80}$$

where now M_s is assumed known for any particular calculation.

I now specify the mass and chemical composition of a star and have a well-defined problem to solve (3.75)–(3.78) with the subsidiary relations (3.70)–(3.72) and the boundary conditions (3.79) and (3.80). In the astronomical literature there is to be found a *theorem* known as the Vogt-Russell theorem which states that, if the mass and chemical composition of the star are specified, this set of equations has only one solution and the structure of the star is uniquely determined. However, no rigorous proof of this theorem was given and it is now known that in some rare circumstances two different solutions of this set of equations may be possible. In that case which solution is relevant must be determined by the past history of the star and by a set of equations in which time dependence occurs. In this book I shall not give any explicit examples of those cases in which the Vogt-Russell theorem is not valid.

Stellar evolution

In this chapter I have obtained a set of equations which determine the structure of a star of given mass and chemical composition and, as these equations contain no time derivatives of the physical quantities, they cannot tell us how the properties of a star change with time. But stars do evolve. They are continually radiating energy into space and thereby losing mass and, as this energy is released by nuclear reactions in the interior of the star, the chemical composition is also changing. This mass loss is slight and cannot exceed 1% of the star's mass in its entire life (although a star might lose mass directly through other causes and this certainly happens in the explosions of novae and supernovae and is observed in many stages of stellar evolution) and this can safely be ignored, but the change in chemical composition is crucial in determining how the properties of a star change as it evolves.

The evolution of a star can only be studied by using equations in which the time derivatives of the physical quantities are included. I have previously argued that the time derivative can be omitted from (3.4) provided that evolution is occurring slowly compared to the dynamical time-scale:

$$t_d = (2r^3/GM)^{1/2}, \tag{3.11}$$

and that will certainly be true for normal stellar evolution. I have also suggested that time dependence can be omitted from (3.47) provided that the series of nuclear reactions at present occurring can supply the star's radiation for a time greatly in excess of the thermal time. That is:

$$t_n \gg t_{th}. \tag{3.48}$$

As we shall see in later chapters of this book there are many stages in stellar evolution when inequality (3.48) is not true and in these stages the evolution can only be calculated by using a more accurate version of (3.47).

However, there are occasions in stellar evolution when inequality (3.48) is satisfied and the set of equations (3.75)–(3.78) is a perfectly adequate set of equations. How then can the evolution of the star be studied? The nuclear reactions which are supplying the energy the star is radiating are changing its chemical composition and I can therefore write down equations describing how this change of composition is occurring.

If there are no bulk motions in the interior of a star, such as would occur if convection were the main mechanism of energy transport, any changes of chemical composition are localised in the element of material in which they are brought about by nuclear reactions. This means that, even if the star started its life with a homogeneous chemical composition, the subsequent composition must be regarded as a function of the mass M. In this simplest case, in which no bulk motions occur, the set of equations must be supplemented by equations describing the rate of change with time of the abundances of the different chemical elements. These equations may be schematically written:

$$\frac{\partial}{\partial t}(\text{composition})_M = f(\rho, T, \text{composition}), \tag{3.81}$$

where the time derivative is taken for a fixed element of mass and where the right-hand side has the form shown because the rate at which nuclear reactions occur depends on ρ, T and composition. In the case when hydrogen is being converted into helium we would have two equations of the form of (3.81). One would describe the decrease in the hydrogen content and the other the increase in the helium content.

Method of solution of the equations

Now I have a time dependent equation (3.81), the derivatives d/dM which occur in (3.75)–(3.78) should strictly be replaced by the partial derivatives $(\partial/\partial M)$, but this does not really matter as there are no equations in which derivatives with respect to both M and t occur and the time and mass dependence of the problem are completely separated. To show what I mean by this I will consider how the evolution of a star could be studied. I first suppose that I know the chemical composition of a star as a function of M at some time t_0; I will not enquire at the moment how I can know this. If the total mass M_s, and the chemical composition as a function of M are known, (3.75)–(3.78) and (3.70)–(3.72) can be solved with the boundary conditions (3.79) (3.80) to determine the complete internal structure of the star at time t_0.

The evaluation of the expression (3.72) for the rate of energy release, ε, has involved a discussion of the rate at which different nuclear reactions are occurring and thus how the chemical composition of the star is changing. The new chemical composition of the star as a function of M at the time $t_0 + \delta t$ can now be obtained from the (schematic) equation:

$$(\text{Composition})_{M,t_0+\delta t} = (\text{Composition})_{M,t_0} + \frac{\partial}{\partial t}(\text{Composition})_M \delta t,$$
(3.82)

where, for example, $(\text{Composition})_{M,t_0}$ denotes the composition at mass M at time t_0. Use of (3.82) at all points in the star gives the chemical composition as a function of M at time $t_0 + \delta t$. With this modified chemical composition (3.75)–(3.78) and (3.70)–(3.72) can again be solved to find the modified structure of the star so that I know how the star's properties have changed between times t_0 and $t_0 + \delta t$. This process can be repeated and the life history of the star can be described.

This process will break down if at any time the properties of the star are found to be changing so rapidly with time that time dependent terms in (3.75)–(3.78) cannot be regarded as unimportant. In that case these equations must be modified. The problem then becomes more difficult mathematically because there are now equations containing derivatives with respect to both M and t and the mass and time dependence of the problem no longer separates as in my discussion above. The equations can, however, be solved and the results of such calculations will be described in later chapters of the book.

Influence of convection

The above discussion is also over-simplified because in most stars there is a region in which a significant amount of energy is carried by convection. Suppose I have calculated the structure of a star of given mass and chemical composition assuming that convection does not occur. I can now check whether that assumption is valid. At all points in the star I have values for P, dP/dM, ρ and $d\rho/dM$ and I can hence calculate $(P d\rho/dM)(\rho dP/dM)$ and compare its value with $1/\gamma$ to see whether or not criterion (3.66) is satisfied anywhere. If it exceeds $1/\gamma$ at all points in the star, convection is indeed absent and the structure of the star has been correctly calculated but, if it is less than $1/\gamma$ anywhere, convection must be occurring and the whole solution of the equations must be reconsidered.

The modification of the equations is as follows. Instead of (3.78) I must use the equation:

$$L_{\text{rad}} = -\frac{64\pi^2 acr^4 T^3}{3\kappa}\frac{dT}{dM},$$
(3.83)

which says that whether or not convection is occurring the amount of energy carried by radiation (which also includes conduction as usual) is determined by the temperature gradient. I can then write:

$$L = L_{rad} + L_{conv} \tag{3.84}$$

and I require an expression for the amount of energy carried by convection, L_{conv}. I write this final equation schematically as:

$$L_{conv} = ?, \tag{3.85}$$

where I indicate that there is no reliable theory which calculates the amount of energy carried by convection and I do not write $L_{conv}(\rho, T, \text{composition})$, because it is not clear that the amount of energy carried by convection across any radius r depends only on conditions at that radius. Such a form will be valid if typical convective elements only travel a short distance, but not if many of them have come from regions with very different physical conditions. In the case of convection in a liquid it is known that the amount of energy carried by convection depends on the depth of the layer of liquid in which convection is occurring.

If I had an expression for the right-hand side of (3.85), the equations of stellar structure would be modified by the replacement of (3.78) by (3.83) to (3.85); of course the expression of (3.85) would ensure that no energy was carried by convection if $P dT/T dP < (\gamma - 1)/\gamma$. It would then still be a reasonably straightforward matter to solve for the structure of a star of given mass and chemical composition. The study of stellar evolution is more difficult because the changes of chemical composition are now not necessarily localised where they occur.

The problem is still reasonably simple if convection only occurs in a region in which there are no nuclear reactions or if the region of nuclear energy release is entirely contained within a convective region. In the former case, (3.81) is still valid. In the latter case the chemical composition of the convective region probably remains essentially uniform as convection currents keep the region well mixed. In the Sun, for example, significant changes of chemical composition cannot at present be occurring in less than about 10^9 years. Convection currents would only have to have a speed of $10^{-7}\mathrm{ms}^{-1}$ to travel across the Sun's radius several times in that time and they would have to be very much faster than that if they were to carry much energy.

I have said that convection will occur if criterion (3.68) is satisfied. Broadly speaking there are two ways in which this can happen. Either, for a gas with a normal value of γ, the temperature gradient required to carry all of the energy by radiation can become large or there may be a region in which the ratio of specific heats becomes close to unity and the criterion can be satisfied for an ordinary value of the temperature gradient. If a large amount of energy is released in a small volume at the centre of a star, it may require a large temperature gradient to carry the energy away. This means that convection may occur in regions in which nuclear energy is released near the centre of stars and such regions are known as convective cores. Specific heats at constant pressure and constant volume approach equality when a change of state is occurring and most of the energy supplied in an attempt to raise the temperature is used up in supplying the latent heat for the change of state. If the surface temperature of a star is low enough for the atoms at the surface to be predominantly neutral, there may be a zone just below the surface in which the abundant elements are being ionised and in which the ratio of specific heats is close to unity. In such a case a star can have an outer convective layer.

Convection in stellar interiors

Although no really adequate expression is available for the right-hand side of (3.85), there are fortunately occasions on which, even though convection occurs, it is possible to avoid using (3.85). In the deep interior of a star in which there is a convective core, it appears that only a very slight increase of the temperature gradient over the adiabatic value defined by

$$\left(\frac{P}{T}\frac{dT}{dP}\right)_{ad} = \frac{\gamma - 1}{\gamma}. \tag{3.86}$$

occurs before convection is capable of carrying *all* of the energy which is required.

An approximate estimate of how much energy can be carried by convection can be made as follows. Heat is convected by rising elements which are hotter than their surroundings and by falling elements which are cooler than their surroundings. Suppose that each type of element has a temperature which differs by δT from that of the surroundings. As a rising element is in pressure balance with its surroundings, it has an energy content per kilogram which is $c_p \delta T$ greater than the energy content of a kilogram of the surrounding medium, where c_p is the specific heat at constant pressure. If I consider the stellar material to be a monatomic perfect gas, c_p is $5k/2m$, where m is the average mass of the particles in the gas. The falling elements have a similar energy deficit. Suppose that a fraction $\alpha(\leq 1)$ of the material is in the rising and falling columns and suppose that they are both moving with a speed v ms^{-1}. Then the rate at which excess energy is carried across a sphere of radius r is:

surface area of sphere × rate of transport of mass

× excess energy per unit mass

$$= 4\pi r^2 . \alpha\rho v . 5k\delta T/2m$$

$$= 10\pi r^2 \alpha v k\delta T/m. \tag{3.87}$$

Fairly near the centre of the Sun I can consider a sphere of radius 10^8m and the density of the material is about 5×10^4 kgm^{-3}.† With k equal to 1.4×10^{-23} JK^{-1} and m taken equal to 8×10^{-28} kg (the value appropriate to fully ionised hydrogen):

$$L_{conv} \simeq 2.5 \times 10^{26} \alpha v \delta T \text{W}. \tag{3.88}$$

---

† This value comes from solutions of the equations of stellar structure but the argument could be made with mean density 1.4×10^3 kgm^{-3}.

At no point in the Sun does the luminosity exceed its surface value of 4×10^{26}W. Thus it can be seen from (3.88) that, provided a reasonable fraction of the material is taking part in the convection, a velocity of a few metres a second and a temperature difference of a few degrees suffice to carry all of the Sun's energy. As the temperature excess and velocity of the rising elements is determined by the difference between the actual temperature gradient and the adiabatic gradient, this suggests that the actual gradient is not greatly in excess of the adiabatic gradient. To a reasonable degree of accuracy I can assume that the temperature gradient has exactly the adiabatic value in a convective core. Thus I take:

$$\frac{P}{T}\frac{dT}{dP} = \frac{\gamma - 1}{\gamma}. \tag{3.89}$$

Note although I have used solar values in this discussion, the same result is valid in other stars. Note also that the Sun does not have a convective core according to theoretical solar models; all I have shown here is that, if it did have a convective core, (3.89) would be almost exactly true in it.

When I have a convective region in which (3.89) is essentially true, I can forget about (3.84) and (3.85) and I can replace (3.83) by (3.89). Thus, in a convective core, the four differential equations

$$\frac{dP}{dM} = -\frac{GM}{4\pi r^4}, \tag{3.75}$$

$$\frac{dr}{dM} = \frac{1}{4\pi r^2 \rho}, \tag{3.76}$$

$$\frac{dL}{dM} = \varepsilon \tag{3.77}$$

and

$$\frac{P}{T}\frac{dT}{dP} = \frac{\gamma - 1}{\gamma} \tag{3.89}$$

must be solved together with equations for ε and P. Equation (3.83) for the radiative flux:

$$L_{\text{rad}} = -\frac{64\pi^2 acr^4 T^3}{3\kappa}\frac{dT}{dM}, \tag{3.83}$$

is of course still true and once the other equations have been solved L_{rad} can be calculated. This can then be compared with L calculated from (3.77) and the difference between the two gives the value of L_{conv}. If convection is occurring, L_{conv} must be positive and if at any time it is found to be negative, this is a signal that the temperature gradient given by (3.89) is more than capable of carrying all of the energy by radiation and that convection will not in fact occur. In an actual star there will possibly be regions in which convection occurs and other regions in which it does not. In solving the equations of stellar structure, the equations appropriate to a convective region must be *switched on* whenever the temperature gradient reaches the adiabatic value and these convective equations must be *switched off* whenever radiation is capable of carrying all of the energy with a temperature gradient lower than the adiabatic value.

Convection near stellar surface

Although, in the central regions of stars, (3.89) is valid if convection occurs, the same is not true near the surface. Thus near the surface of the Sun, where r is 7×10^8 m and $\rho \simeq 10^{-3}$ kgm^{-3},

$$L_{\text{conv}} \simeq 2.5 \times 10^{20} \alpha v \delta T \text{W}. \tag{3.90}$$

In this region the temperature is about 10^4K and the velocity of sound 10^4ms^{-1}. It is clear that, in order for the convective luminosity to be comparable with the total energy transport of 4×10^{26}W, the velocity of rising and falling elements must be comparable with the velocity of sound and the temperature differences must be comparable with the actual temperature. Such velocities and temperature differences can only be driven by a temperature gradient substantially in excess of the adiabatic value and in such a case an expression for the amount of energy carried by convection (3.85) is required. It is in such low density surface regions that the lack of a good theory of convection can be serious. In the case of the Sun, the convection is observed directly in the solar granulation.

The mixing length theory of convection

The description of convection which is commonly used in stellar interiors contains a free parameter which is called the *mixing length*, l. It is assumed that convective elements of a characteristic size l rise or fall through a distance which is comparable with their size before exchanging their excess heat with their surroundings. If it is assumed that the elements move adiabatically and in pressure balance with their surroundings and that they are accelerated freely by the bouyancy force, an expression for the convective luminosity can be obtained in the form

$$L_{\text{conv}} = \pi r^2 c_p \rho T (GM/r^2)^{1/2} l^2 (\nabla - \nabla_{\text{ad}})^{3/2} / H_p^{3/2}, \tag{3.91}$$

where

$$H_p = P/(-dP/dr) \tag{3.92}$$

is called the *pressure scale height*,

$$\nabla = PdT/TdP, \tag{3.93}$$

$$\nabla_{\text{ad}} = (\gamma - 1)/\gamma, \tag{3.94}$$

and c_p is the specific heat at constant pressure.

The expression (3.91) is only useful if a value can be chosen for l. It is generally agreed that an appropriate distance for convective elements to travel is of the order of a pressure scale height, but it makes a considerable difference to the calculated properties of some cool stars if $\alpha(\equiv l/H_p)$ is taken to be 0.5 rather than 1.5 say. In the case of the Sun, so much is known in detail about its properties that a value of α can be chosen which provides the best match between theory and observation. There is then a temptation to use the same value of α for other stars but there is really no justification for doing this. An improved theory of convection is still very badly needed.

It can be seen that the value for L_{conv} in the mixing length theory depends on the (3/2)th power of the difference between PdT/TdP and $(\gamma - 1)/\gamma$. If this expression is used in convective cores where I have previously assumed that this difference is zero, it is possible to estimate what the difference really is. In the case of massive stars, which have a significant convective core, it is found that the difference is of order 10^{-6} so that the approximation of taking it to be zero is very good indeed.

Convective overshooting

There is one further important property of convection. In my discussion so far I have assumed that the boundary of any convection zone occurs where $PdT/TdP = (\gamma - 1)/\gamma$. In fact, if a convective element rises through a region in which the criterion for instability, (3.68), is satisfied, it will still be travelling with a finite velocity when it enters the region in which the criterion is not satisfied. Although it will now be retarded it will not stop instantaneously. This means that convection will extend somewhat further than I have assumed. This process is called *convective overshooting*. This is probably not of great importance for energy transport but it may affect the mixing of material in stellar interiors. As I shall explain in Chapter 6, as a star evolves it may possess both internal and external convection zones which increase and decrease in size. If material, which has at one stage been in an internal convection zone in which nuclear reactions are occurring, is subsequently in an outer convection zone, nuclearly processed material can appear at the surface of a star. The extent of convective overshooting that occurs may determine whether such an overlap of convection zones occurs. Overshooting may also play a rôle in the evolution of very massive stars which have large convective cores.

Complications in real stars

In this Chapter I have discussed the equations appropriate to the structure of spherically symmetrical stars of constant mass. If these equations are solved for the evolution of a star, at each stage the equilibrium stellar model ought to be tested to determine whether or not it is stable. The only instability which I have discussed is a local one which leads to convection but there are other instabilities which may lead to stellar mass loss and explosion, with the most violent event being the explosion of a supernova. Other instabilities are stabilised at finite amplitude and they lead to the observed variable stars. In Chapter 2 I have explained that some regions in the HR diagram are populated by variable stars. It is the task of theories of stellar structure to explain why stars with these observed surface properties are unstable. The assumption of constant mass is now known to be a much less good assumption than was once thought and I shall discuss mass loss from stars in Chapter 7. Although some mass loss results from a violent instability, in most cases of observed mass loss there just is not an equilibrium solution for a star of constant mass when the properties of its outer layers are treated carefully.

Departures from spherical symmetry are also important in some stars. There are three obvious causes of non-sphericity, rapid rotation, a strong internal magnetic field which cannot exert a spherically symmetric force, and the gravitational attraction of a close companion star, which is usually accompanied by rapid rotation. The evolution of close binary stars is particularly important because the growth in size of one of the components as a result of normal stellar evolution can lead to matter being transferred from one star to another. When the transfer occurs to a very compact companion star, the white dwarfs and neutron stars which I shall discuss in Chapter 10, very dramatic results can occur. It is now believed that all novae and some supernovae involve mass transfer in binary stars with compact components. I shall discuss close binary stars in Chapter 8.

Summary of Chapter 3

In this chapter the equations of stellar structure have been formulated. Stars are held together by the force of gravitation and the gravitational force on unit volume is resisted by the pressure gradient of the stellar material. Because the properties of most stars are slowly varying, these forces must be almost exactly in balance and from this it is possible to deduce that the temperatures inside stars are higher than 10^6 K and that, despite stellar densities being as high as solids, the stellar material is gaseous. It is, in fact, a gas of ions and electrons rather than atoms and molecules. If it is an ideal classical gas, it follows from the Virial Theorem that a star becomes hotter as it loses energy.

The luminosity of the Sun has not changed significantly in the past 10^9 years and only energy released by nuclear fusion reactions can have supplied all of this energy. Nuclear energy is released at the hottest regions in a star near its centre and it must be carried to the surface by conduction, convection or radiation. The transport of energy by conduction and radiation depends on reasonably well-understood physical processes and it appears that conduction is unimportant in most stars. The condition for the onset of convection is clear, but there is at present no good theory which predicts how much energy is carried by fully developed convection. Fortunately, in many cases, convection can be shown to be so efficient that it will carry all of the energy required and then a detailed theory of convection is not needed.

The calculation of the structure of a star requires the specification of its mass and chemical composition and the solution of four differential equations with two boundary conditions at both the centre and the surface. In many stages of a star's evolution, all time derivatives are small and can be omitted from the equations. It is then relatively simple to study the slow evolution of a star. The structure of the star can first be calculated at a given time. Nuclear reactions then cause a gradual change in its chemical composition and the structure can be recalculated a short time with the revised chemical composition. This procedure can then be repeated.

4

The physics of stellar interiors

Introduction

In the last chapter I have derived the equations of stellar structure. In these equations there are three quantities, the pressure, P, the energy release per kilogram per second, ε, and the opacity coefficient, κ, which depend on the density, temperature and chemical composition of the stellar material. In Chapter 3 I did not discuss how P, ε and κ depend on these quantities, and the present chapter is concerned with discussing how P, ε and κ are to be calculated if the density, temperature and chemical composition are known. To calculate ε a considerable knowledge of nuclear physics is required and a similar knowledge of atomic physics is required for the determination of κ. All three quantities depend on the thermodynamic state of the stellar material. Once ρ, T and the chemical composition are known, the calculation of P, ε and κ is pure physics and no further astronomical concepts are required and it is for this reason that this chapter is called *The physics of stellar interiors*.

Because of the great complexity of the problems involved I am only able to describe the basic processes determining P, ε and κ and am not able to give detailed calculations. In the first place, I consider the law of energy release.

Energy release from nuclear reactions

As mentioned in the last chapter, it is now believed that most of the energy radiated by stars has been released by nuclear reactions in the stellar interior. I first consider why it is that energy can be released by nuclear reactions and how it is determined which nuclear reactions will release energy.

Atomic nuclei are composed of protons and neutrons, which together are referred to as nucleons. The total mass of a nucleus is less than the mass of its constituent nucleons. That means that there is a decrease in mass if a compound nucleus is formed from nucleons and by the Einstein mass–energy relation, $E =$

80

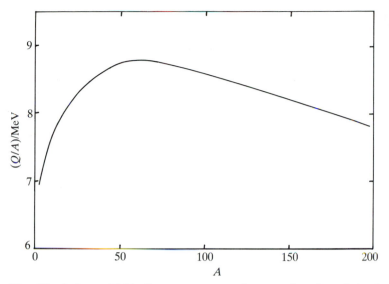

Fig. 37. The (schematic) binding energy per nucleon as a function of atomic mass; Q is measured in MeV. The true curve is much more irregular but has the overall properties shown.

mc^2, the lost mass is released as energy. This energy is known as the binding energy of the compound nucleus and this can be calculated when the mass difference between the compound nucleus and its constituent nucleons is known. Thus, if a nucleus is composed of Z protons and N neutrons, its binding energy $Q(Z, N)$ is :

$$Q(Z, N) \equiv [Zm_{\mathrm{p}} + Nm_{\mathrm{n}} - m(Z, N)]c^2, \qquad (4.1)$$

where m_{p} is the proton mass, m_{n} the neutron mass and $m(Z, N)$ the mass of the compound nucleus.

A more significant quantity for our present discussion than the total binding energy of a nucleus is the binding energy per nucleon $Q/(Z + N)$. This is proportional to the fractional loss of mass when the compound nucleus is formed. If $Q(Z, N)/(Z + N)$ is plotted against A ($\equiv Z + N$) for a large number of nuclei, the resulting diagram has the general character shown in fig. 37. The actual curve is very irregular; in particular for a given value of A there may be several isobars (nuclei which have the same number of nucleons, but a different division between protons and neutrons) which have different values of Q. The general property of the curve is that the binding energy per nucleon rises rapidly with the initial increase in nucleon number, there is a broad maximum for A values between 50 and 60 (that is for nuclei in the neighbourhood of iron in the periodic table) and then there is a gradual decline for nuclei with higher values of A. The nuclei in the neighbourhood of iron are the most strongly bound nuclei. They have the greatest fractional loss of mass when formed out of individual protons and neutrons.

From fig. 37 the possibility of nuclear fusion and fission reactions releasing energy can be deduced. Thus consider first the process by which light nuclei combine (fuse) to form heavier nuclei. If the two nuclei and their compound all lie to the left of the maximum in fig. 37, the compound nucleus has a larger binding energy per nucleon than the original nuclei and, as the total number of nucleons has not been changed, the nuclear reaction must release energy. An example of such an energy releasing fusion reaction is the combination of a helium nucleus and a carbon nucleus to produce an oxygen nucleus :

$$^4\text{He} + {}^{12}\text{C} \rightarrow {}^{16}\text{O} + 7.2\,\text{MeV}. \tag{4.2}$$

If a succession of fusion reactions takes place in which no more than a few nucleons are added at any time, further energy release from fusion reactions becomes impossible when the nucleus is in the iron region.

Fission reactions, in which a heavy nucleus splits into two or more fragments, will also clearly release energy if all of the nuclei involved are to the right of the maximum in fig. 37. In this case also the newly formed nuclei have a larger binding energy per nucleon than the initial nucleus so that it is clear that energy has been released by the reaction.† In the case of some of the heaviest nuclei these fission reactions occur spontaneously as in the fission of ^{235}U.

Fission reactions provide the energy release in the atomic bomb and in nuclear reactors. The fusion of light nuclei provides the energy of the hydrogen bomb. In addition since about 1950 scientists in many countries have been trying to produce nuclear fusion reactions between the heavy isotopes of hydrogen, deuterium and tritium, under controlled conditions in the laboratory. The problems of producing controlled thermonuclear reactions appear to be very difficult because the material must be raised to a temperature in excess of 10^8K and kept at that temperature long enough to produce a useful yield of energy.* At the same time the hot reacting material must be insulated from the walls of the containing vessel. If these difficulties are eventually overcome, the world's useable energy resources will be vastly increased.

The energy release from fusion reactions converting hydrogen into the most abundant isotopes of helium (^4He) and iron (^{56}Fe) is, in $\text{J}\,\text{kg}^{-1}$,

$$\left.\begin{array}{lll} \text{H} \rightarrow {}^4\text{He}, & 6.3 \times 10^{14} \\ \text{H} \rightarrow {}^{56}\text{Fe}, & 7.6 \times 10^{14}. \end{array}\right\} \tag{4.3}$$

The latter is about the maximum energy release which can possibly be obtained from nuclear fusion reactions and, since the rest mass energy of 1 kg is 9×10^{16} J, this maximum energy release is just under 1% of the rest mass energy as stated in Chapter 3. The binding energy per nucleon of very heavy nuclei, although less than that of iron, is still quite large. This means that the maximum possible energy release per kg from fission reactions is much less than that from fusion reactions.

---

† In actually occurring fission reactions some of the fragments may lie to the left of the maximum in the binding energy curve and a more careful discussion is required to demonstrate the energy release.

* The temperature requred for self-sustaining laboratory fusion reactors is higher than that in stars because the laboratory plasma is not opaque to radiation which can escape from the vessel, whereas stars radiate from a surface which is much cooler than the interior.

Thus, if there were comparable amounts of heavy and light elements in stars, we would expect fusion reactions to be a more important source of energy than fission reactions. In fact as we have seen in fig. 21 in Chapter 2, the very heavy nuclei do not appear to be very abundant in nature and it is thus believed that nuclear fusion reactions are by far the main source of the energy radiated by stars.

The four forces of physics

From fig. 37 we have seen that nuclear fusion reactions are energetically possible but I must now discuss the conditions under which they will actually occur and decide whether these conditions will be found in stars. For example, it is obviously not true that hydrogen changes spontaneously into iron under all conditions. In stellar interiors the material is highly ionised and I am interested in reactions between bare nuclei. Nuclei interact through the four basic interactions of physics : electromagnetic, gravitational, strong nuclear and weak nuclear. Of these the two most important in the present subject are the electromagnetic and strong nuclear interactions. Gravitational forces are vitally important for the structure of whole stars, but are completely negligible in interactions between individual particles. A measure of the weakness of the gravitational interactions is the ratio of the gravitational force between a proton and an electron to the electrostatic force between the same particles. This is :

$$4\pi\varepsilon_0 G m_p m_e / e^2 \simeq 4 \times 10^{-40}, \tag{4.4}$$

where m_e and e are the mass and charge of the electron respectively and ε_0 is the permittivity of free space.

As the electrostatic and gravitational interactions are both inverse square forces such a comparison is easily made. The strong and weak nuclear interactions do not share this property as they are both short range interactions which means that they are only important when particles are very close together. In the region where they are important, the strong nuclear interaction is much stronger than the electrostatic interaction, while the weak nuclear interaction, although weaker than the electrostatic interaction, is very much stronger than gravitational forces. The particular importance of the weak interaction is that, if it did not exist, many unstable elementary particles, such as the neutron, would be stable. Thus neutron decay

$$n \rightarrow p + e^- + \bar{\nu}_e \tag{4.5}$$

in which a neutron (n) is converted into a proton (p), an electron (e⁻) and an anti-neutrino† ($\bar{\nu}_e$) would not occur if it were not for the existence of the weak

---

† Every elementary particle possesses an anti-particle which has equal mass but equal and opposite values for other basic properties such as electric charge. For example, the positron is the anti-particle of the electron. There are two neutrinos, one of which is called the neutrino and the other the anti-neutrino. When a positron is emitted in β-decay it is accompanied by a neutrino while an electron is accompanied by an anti-neutrino.

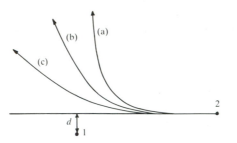

Fig. 38.   Effect of the electric charges of two nuclei, 1 and 2, on their relative motion.

nuclear interaction. I shall explain that a similar reaction converting a proton into a neutron plays an important part in the conversion of hydrogen into helium as in that conversion two protons must be changed into neutrons.

I first consider only the electromagnetic and strong nuclear interactions between nuclei. The strong nuclear interaction has a very short range and two nucleons only affect one another through it if they are separated by less than about 10^{-15}m. At this separation and rather less, the interaction is attractive and it is the force which holds nuclei together and which causes the nuclear reactions which occur in stars. Thus if two nuclei come close enough to be influenced by it, they are drawn together and may form a compound nucleus.

The electrostatic force between two positively charged nuclei is repulsive and, unlike the strong nuclear interaction it has a long range, only falling off with distance, r, as $1/r^2$. The existence of stable nuclei depends on the balance between these two forces. Because all of the protons in a nucleus repel all of the other protons through their electrostatic interaction and because, in contrast, only the nearby protons and neutrons strongly attract one another through the nuclear force, in heavy nuclei the proportion of nucleons in the form of protons decreases. If there were more protons, the repulsive electrostatic force would be too strong to allow the nucleus to exist. In the case of ^{238}U for example there are 146 neutrons and only 92 protons.

Occurrence of nuclear reactions

If I consider not the structure of a given nucleus but the possibility that a nuclear interaction between two nuclei can lead to a nuclear transmutation, I see that the repulsive electromagnetic interaction between two postively charged nuclei tends to prevent them from approaching close enough for the strong nuclear interaction to become significant. Nuclear reactions can occur if particles approach close enough for this to happen despite the repulsive effect of the electric charges and this occurs if the relative velocity of the two particles is high enough.

This is illustrated in fig. 38. A coordinate system is chosen in which one particle is at rest. The other particle is moving in a direction which, in the absence of the electrostatic force, would lead to its having a distance of closest approach d to particle 1. The actual track of particle 2 is a hyperbola and curves (a), (b) and (c) show a sequence of tracks for increasing initial velocity of particle 2. The higher this velocity the closer is the distance of closest approach to d. If d is less than the range of nuclear forces and the velocity of particle 2 is high enough, there is then a possibility that a nuclear reaction will occur.

This classical description of the occurrence of nuclear reactions has been modified in one very important respect by the quantum theory. This states that there is a possibility that nuclear reactions will occur even when classical theory predicts that particles do not approach close enough for an interaction. This arises through Heisenberg's principle of uncertainty which states that it is impossible to give a precise value to both the position and the momentum of a particle. There is an uncertainty δx in the position x and an uncertainty δp in the momentum p and these are related by:

$$\delta x \delta p \geq h/4\pi, \tag{4.6}$$

where h is Planck's constant. In a similar way Heisenberg's principle states that the energy of a particle cannot be measured in such a way that it and the time of measurement are known precisely. Thus there are uncertainties δE and δt in energy and time such that

$$\delta E \delta t \geq h/4\pi \tag{4.7}$$

If the extra energy above the classical value that an incoming particle requires to approach within the range of nuclear forces and the time for which it requires that energy satisfy (4.7), there is a chance of a nuclear reaction occurring.

As a result the rate at which nuclear reactions occur is much greater than would be expected on a classical theory. Before the quantum theory of nuclear reactions had been developed, it was difficult to see how nuclear reactions could produce the amount of energy which stars were radiating at the temperatures estimated in stellar interiors. At the same time it was difficult to envisage a satisfactory alternative source of energy, although the possibility of complete annihilation of matter into energy was considered. It was at this time that Eddington made his famous remark that, if the centre of the Sun was not hot enough for the nuclear physicists, they must find a hotter place.

Inside a star the velocities possessed by particles are those of their random thermal motions. According to the kinetic theory of gases, the mean speed of a particle, $\bar{v}$, is given by the relation :

$$1.086\bar{v} = (3kT/m)^{1/2} \text{ m s}^{-1} \tag{4.8}$$

where as before m is the mean mass of the particles in the gas. This means that, if any of the particles are to have high enough velocities for nuclear reactions to occur, the temperature of the stellar gas must be high. The reactions are then known as *thermonuclear reactions*. We have already seen in the last chapter that, while the stellar gas remains ideal, its temperature will rise as the star loses energy (see the discussion on page 60). Thus it may be expected that, if the interior of a star is initially not at a high enough temperature for nuclear reactions to occur releasing energy, it will eventually become hot enough. At that stage what I have called hidden energy supplies on page 60 become available and the star can settle into a quasi-steady state in which the energy supplied by thermonuclear reactions supplies the surface energy loss.

The higher the electric charges of interacting nuclei, the greater is the repulsive force between them and the higher the temperature of the stellar gas must be before thermonuclear reactions occur. The highly charged nuclei are also the more massive nuclei and this means that nuclear reactions between light elements occur at lower temperatures than nuclear reactions between heavy elements. Thus in an individual star it may be expected that the light elements will gradually be converted into heavier elements as the star evolves and its internal temperature rises, until finally the material has been converted into elements in the neighbourhood of iron in the periodic table. Once this has happened no more energy can be released by nuclear fusion reactions. It is possible that fusion reactions will not go this far. The proof in Chapter 3 that the stellar central temperature will rise as the star loses energy depends on the material of the star remaining an ideal classical gas. I shall show later in this chapter that, if the stellar gas is influenced by quantum effects, it may be possible for the star's temperature to pass through a maximum and for the star to cool down and *die*. It is also possible that a star will reach its final state as a result of substantial mass loss. The final stages in the evolution of such stars will be discussed in Chapter 10.

As well as depending on the temperature of the stellar material, the rate of nuclear reactions clearly also depends on the density, but in this case the dependence is very simple. Thus, for the simplest two particle nuclear reactions in which two nuclei combine to form a third nucleus, probably with the emission of a photon, the energy release per unit volume is proportional to the product of the numbers of the two interacting particles in unit volume. This is, if nucleus A combines with nucleus B to form nucleus C and a photon γ through the reaction

$$A + B \rightarrow C + \gamma, \dagger \tag{4.9}$$

the number of reactions occurring is proportional to $n(A)n(B)$, where $n(A), n(B)$ are the numbers of particles of types A and B in unit volume. If the chemical composition is fixed, this means that the rate of energy generation per unit volume is proportional to ρ^2 and the rate of energy generation per unit mass, ε, is thus proportional to ρ. These two particle nuclear reactions are usually more important in stellar interiors than reactions involving three or more particles. This is true because the probability of more than two particles being simultaneously close enough together for a reaction involving all of them to take place is very small indeed, unless the density of the material is extremely high. It will, however, soon be seen that there is one very important three particle reaction which leads to an energy release proportional to ρ^2.

---

† In what follows I shall sometimes use an abbreviated notation for such reactions, $A(B,\gamma)C$, where the initially present particles are written to the left of the comma and the final particles to the right. The notation γ is used for a photon as in these nuclear reactions the photons are γ-rays.

Nuclear reaction rates

From what I have said earlier it should be clear that the probability of a nuclear reaction occurring can be written as a product of two factors. These are the probability of two particles approaching close enough for the nuclear force to be important and the probability that a nuclear reaction will then occur. The first factor depends only on the masses and charges of the two particles, the number of particles present and the temperature. This factor is easy to calculate in principle, but requires a knowledge of quantum mechanics beyond the scope of this book. The second factor depends on the detailed properties of the two nuclei involved. It is not usually possible to calculate this factor and it must be determined from laboratory experiments.

Although I cannot calculate the rate at which thermonuclear reactions occur, it is perhaps useful to write down the formula to show how the reaction rate depends on T and the masses and charges of the particles. Suppose two interacting particles have masses $A_i m_H$ and $A_j m_H$, where m_H is the mass of the hydrogen atom and charges $q_i e$ and $q_j e$, where e is the electron charge. Suppose also that a fraction X_i by mass of the material is in the form of nucleus i and a fraction X_j in the form of nucleus j. Define the two quantities:

$$A = A_i A_j / (A_i + A_j) \tag{4.10}$$

and

$$\tau = 4.25 \times 10^3 (q_i^2 q_j^2 A / T)^{1/3}. \tag{4.11}$$

Then the number of reactions per kg per s involving the nuclei i and j can be written

$$R_{ij} = C\rho \frac{X_i X_j}{A_i A_j} \tau^2 \exp(-\tau)(A q_i q_j)^{-1}, \tag{4.12}$$

where C is a constant depending on the particular properties of the nuclei concerned.

In (4.12) the dependence of the reaction rate on density, temperature, nuclear abundances and nuclear masses and charges is shown clearly. When T is small, τ is large and the term $\exp(-\tau)$ leads to a very small reaction rate. As τ decreases, the reaction rate increases rapidly through the exponential term but this increase does not continue for ever. Eventually the term in τ^2 becomes more important than the exponential term when the temperature is very high and the reaction rate drops again. In practice we find that we are interested in temperatures at which there is still a rising trend in the reaction rate. The decrease in reaction rate as the charges on the interacting nuclei are increased is also apparent from (4.11) and (4.12) as there is a strong dependence of q_i and q_j in the exponential term.

Resonant nuclear reactions

Expression (4.12) gives the rate of what are known as non-resonant nuclear reactions. Atomic nuclei just like atoms have a series of energy levels and the probability of a nuclear reaction occurring is greatly increased if the energy available in a nuclear collision is just sufficient to put a compound nucleus into one of its energy levels. This is known as a resonance. For most energies reactions are non-resonant and that is the case for the reactions converting hydrogen to helium which I will now describe. A resonance is however crucially important in the subsequent conversion of helium to carbon.

Hydrogen burning reactions

It is now believed that the most important series of reactions occurring in stars are those converting hydrogen into helium. This is known as *hydrogen burning*. An important feature of these reactions is that they involve the conversion of two protons into neutrons for each nucleus of ^{4}He (α particle) produced. As was mentioned on page 84 the conversion of a proton into a neutron requires the operation of the weak nuclear interaction. Thus hydrogen burning involves rather more than two particle nuclear reactions of the type which has just been described. Two basic reaction chains have been proposed: the proton–proton (PP) chain and the carbon–nitrogen (CN) cycle.

In the first chain hydrogen is converted directly into helium, but in the CN cycle nuclei of carbon and nitrogen are used as catalysts.

Details of the PP chain are shown in (4.13)–(4.15) and of the CN cycle in (4.16). The notation used has been explained in the footnote on page 86. The proton–proton chain divides into three main branches which are called the PP I, PP II and PP III chains. The first reaction is the interaction of two protons to form a

PP I Chain

$$
\left.
\begin{array}{ll}
1 & p(p, e^+ + \nu_e)\, d \\
2 & d(p, \gamma)^3He \\
3 & {}^3He(^3He, p + p)^4He
\end{array}
\right\} \tag{4.13}
$$

nucleus of heavy hydrogen (deuteron, d) with the emission of a positron (e^+) and a neutrino (ν_e). The deuteron then captures another proton and forms the light isotope of helium with the emission of a γ-ray. At this stage two important possibilities arise. The nucleus of ^{3}He can either interact with another nucleus of ^{3}He or with an α particle, which has already been formed, or may have been present initially if the star contained helium at birth. Present views on the origin of the elements suggest that this will always have been true.

In the former case we have the final reaction of the PP I chain while the latter reaction leads into either the PP II or the PP III chain. The remainder of the reactions will not be described in detail, but it should be noted that there is another choice in the chain when ^{7}Be either captures an electron to form ^{7}Li or captures another proton to form ^{8}B. At the end of the PP III chain, the unstable nucleus ^{8}Be breaks up to form two α particles.

PP II Chain

This starts with reactions 1 and 2.

$$\left.\begin{array}{ll} \text{Then } 3' & {}^3\text{He}({}^4\text{He}, \gamma){}^7\text{Be} \\ 4' & {}^7\text{Be}(\text{e}^-, \nu_\text{e}){}^7\text{Li} \\ 5' & {}^7\text{Li}(\text{p}, \alpha){}^4\text{He} \end{array}\right\} \tag{4.14}$$

PP III Chain

This starts with reactions 1, 2 and 3'.

$$\left.\begin{array}{ll} \text{Then } 4'' & {}^7\text{Be}(\text{p}, \gamma){}^8\text{B} \\ 5'' & {}^8\text{B}(, \text{e}^+ + \nu_\text{e}){}^8\text{Be} \\ \text{and } 6'' & {}^8\text{Be} \rightarrow 2{}^4\text{He} \end{array}\right\} \tag{4.15}$$

CN Cycle

$$\left.\begin{array}{ll} 1 & {}^{12}\text{C}(\text{p}, \gamma){}^{13}\text{N} \\ 2 & {}^{13}\text{N}(, \text{e}^+ + \nu_\text{e}){}^{13}\text{C} \\ 3 & {}^{13}\text{C}(\text{p}, \gamma){}^{14}\text{N} \\ 4 & {}^{14}\text{N}(\text{p}, \gamma){}^{15}\text{O} \\ 5 & {}^{15}\text{O}(, \text{e}^+ + \nu_\text{e}){}^{15}\text{N} \\ 6 & {}^{15}\text{N}(\text{p}, {}^4\text{He}){}^{12}\text{C} \end{array}\right\} \tag{4.16}$$

The important feature of the CN cycle is that it starts with a carbon nucleus to which are added four protons successively. In two cases the proton addition is followed immediately by a β-decay, with the emission of a positron and a neutrino, and at the end of the cycle a helium nucleus is emitted and a nucleus of carbon remains. The carbon and nitrogen act purely as catalysts in these reactions and are neither produced nor destroyed. Actually in both the PP chain and the CN cycle there are some less important side reactions which have not been listed. If, of course, any stars exist which do not contain any carbon or nitrogen, the CN cycle cannot occur and all hydrogen burning must be through the PP chain. However, a very small amount of carbon will suffice to make the CN cycle important in some stars as we shall see when I discuss how the energy release from the two reaction chains depends on ρ, T and chemical composition.

Carbon and nitrogen are described as catalysts in the CN cycle because they are not destroyed by its operation. At first sight one might expect that the abundance of the different stable isotopes is not changed by the operation of the cycle but this is not true. When the cycle is working in equilibrium the rates of all the reactions in the chain must be the same. In order for this to be so the abundances of the isotopes must take up values so that those isotopes which react more slowly have the higher abundances. The slowest reaction in the chain is the capture of a proton by ${}^{14}\text{N}$. As a result most of the ${}^{12}\text{C}$ is converted to ${}^{14}\text{N}$ before the cycle reaches equilibrium. In addition the amount of ^{13}C relative to ^{12}C increases. The abundance ratios ^{13}C/^{12}C and ^{14}N/^{12}C when the CN cycle is fully operational are both very much higher than is found in solar system material. Abundance ratios comparable with the CN cycle values are observed in some stars and this is taken as good evidence both that the CN cycle has operated in the stars and that the reaction products have been carried to the stellar surfaces. This will be mentioned further in Chapter 9.

Neutrinos

These hydrogen burning reactions are of particular interest because the conversion of a proton into a neutron involves the emission of not only a positron but also a neutrino. The neutrino is a particle which is apparently without mass, possesses no electromagnetic properties (charge, magnetic moment etc.) and does not take part in strong interactions. In fact it was originally suggested that it must exist because otherwise energy and momentum would not be conserved in β-decay reactions. Rather than jettison apparently well-established conservation laws, it was felt preferable to hypothesize that an additional, as yet undetected, particle was emitted in the reaction.

After more than 20 years the neutrino (or more accurately the anti-neutrino) was detected in 1953. Neutrinos interact very weakly with other matter and a neutrino of 1 MeV energy would pass through about 10 parsecs of water without being seriously deflected or absorbed. It is thus very difficult to detect an individual neutrino but, since there is a small, but finite, probability that a neutrino will be absorbed in a short distance, neutrinos can be detected provided a sufficiently strong flux of neutrinos can be obtained. In the first experiments the anti-neutrinos were emitted by β-decay of unstable nuclei; inside the nucleus an individual neutron decayed through reaction (4.5):

$$n \rightarrow p + e^- + \bar{\nu}_e,$$

The anti-neutrinos were then captured by protons through a reaction which is essentially the inverse of (4.5):

$$\bar{\nu}_e + p \rightarrow n + e^+. \tag{4.17}$$

Subsequently the neutron decayed through reaction (4.5) and the positron annihilated with an electron to produce γ-rays:

$$e^+ + e^- \rightarrow \gamma + \gamma. \tag{4.18}$$

The neutron decay and the gamma ray production were both observed and the near coincidence of these events enabled occurrence of the anti-neutrino capture reaction to be deduced. It is hardly surprising that the positive detection of the anti-neutrino was not obtained easily.

Almost all of the neutrinos emitted by nuclear reactions in the centre of a star escape from the star without any further interaction, whereas, as will be seen later in this chapter, the low energy γ-rays emitted in the nuclear reactions only travel a small fraction of the star's radius before they are absorbed. The neutrinos in reactions (4.13)–(4.16) carry away between 2% and 6% of the energy released in the reactions and this energy is lost almost immediately and is of no use to the star. Thus the period of hydrogen burning is a few per cent shorter than it would otherwise have been. Perhaps the most interesting property of these neutrinos is that they are potentially capable of giving the observer on Earth some information about conditions in the centre of the Sun whereas photons only give direct information about the surface layers. Although neutrinos interact only very weakly with matter there is a possibility that a few can be detected and can give some direct information about the region in which nuclear reactions are occurring in the Sun.

Since the mid-1960s an experiment has been under way to try to detect these neutrinos in a large tank containing 400 000 litres of cleaning fluid (perchloroethylene C_2Cl_4) placed almost a mile underground in a gold mine! The neutrinos are absorbed by ^{37}Cl, the heavy isotope of chlorine, through the reaction

$$^{37}Cl + \nu_e \rightarrow {}^{37}Ar + e^-. \tag{4.19}$$

The radioactive argon is then separated by a chemical process from the remainder of the fluid and the number of radioactive atoms produced is counted by observing the reverse reaction:

$$^{37}Ar \rightarrow {}^{37}Cl + e^+ + \nu_e. \tag{4.20}$$

The experiment is placed deep in a mine to minimize the production of ^{37}Ar by other agents such as cosmic rays. It is possible to calculate how many neutrinos should be detected in the experiment but ever since it has been running fewer neutrinos have been detected than were expected. The discrepancy is a factor of order 3. Many attempts have been made to understand this result without there being any agreed solution to date. Recently a second experiment has come into operation using gallium as the absorber of the neutrinos. The results of this experiment also appear to be different from theoretical predictions. I shall discuss the experiments further in Chapter 6.

Energy release from hydrogen burning

Using experimental, or in many cases, extrapolated values for the rates at which the reactions in the chains (4.13)–(4.16) occur, it is possible to tabulate the energy release of the CN cycle and the total release from the three branches of the PP chain as a function of temperature and this is shown in fig. 39. In both cases the energy release per kg is proportional to the density. The energy release from the PP chain is proportional to the square of the fractional hydrogen content X_H while that for the CN cycle is proportional to the product of the hydrogen concentration and the carbon concentration X_C. The diagram is drawn for a value of X_C appropriate to population I stars like the Sun which are relatively rich in elements heavier than hydrogen and helium. A change in the value of X_C/X_H only leads to a bodily displacement of one curve relative to the other.

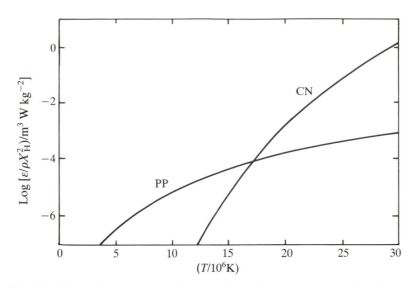

Fig. 39.   Rate of energy release from hydrogen burning as a function of temperature for the two reaction chains.

Consider the implication of this diagram for the Sun. The Sun's energy release is at the rate of 2×10^{-4} W kg^{-1} averaged over its whole mass. The mean density of the Sun is 1.4×10^{3} kgm^{-3} and the observations of the chemical composition of the solar surface suggest that $X_H \simeq \frac{3}{4}$. If the entire Sun were at the same temperature and density, it would have a value of $\varepsilon/\rho X_H{}^2$ of about 2.5×10^{-7} and it can be seen from the figure that the temperature would be about 5×10^6K. In fact the flow of heat outwards from the centre of the Sun requires the central temperature to be higher than the mean temperature so that this estimate is certainly somewhat low. This estimate of the interior temperature of the Sun can be compared with a lower limit to the mean temperature of 2×10^6K which was obtained from (3.32). Although the estimate I have made here is very rough, it is satisfactory that hydrogen burning reactions will supply energy at the observed rate for stellar interior temperatures which are quite similar to those estimated in Chapter 3.

The energy released by the PP chain and CN cycle are smooth functions of temperature. In a limited temperature range, I can replace the true dependence on temperature shown in fig. 39 or (4.11) by a power law obtained by replacing the curve by its tangent, or perhaps better by an appropriate chord parallel to the tangent at some point in the range. Thus, in a region not far below the temperature where the PP chain and CN cycle are equally important, I can represent the energy generation rate by:

$$\varepsilon_{pp} \simeq \varepsilon_1 X_H^2 \rho T^4, \tag{4.21}$$

while for the CN cycle at a slightly higher temperature:

$$\varepsilon_{\mathrm{CN}} \simeq \varepsilon_2 X_{\mathrm{H}} X_{\mathrm{C}} \rho T^{17}, \tag{4.22}$$

where in (4.20) and (4.21) ε_1 and ε_2 are constants.

Clearly the true law of energy release is not a power law, but this is quite a good approximation as the energy release increases very rapidly with temperature and the range of temperature in which significant release occurs is small. In the next chapter I shall show that the use of approximate laws of energy production of the form

$$\varepsilon = \varepsilon_0 \rho T^\eta \tag{4.23}$$

enables us to obtain useful qualitative information about the structure of stars.

Other nuclear reactions involving light elements

It has been stated several times earlier that stars composed of an ideal classical gas heat up as they radiate and that this heating up proceeds until thermonuclear reactions start in the interior. It has also been stated that reactions involving the nuclei of lowest charge occur first so that it might be expected that no significant nuclear reactions can occur at a lower temperature than that at which hydrogen burns to helium. This is not quite true because both the PP chain and CN cycle are atypical. In each case weak interactions must occur to change protons into neutrons and these weak interactions are sometimes slower than strong interactions, although as I have already mentioned the slowest reaction in the CN cycle is a proton capture. In addition, each chain has another peculiarity. The first reaction of the PP chain really involves two steps:

$$p + p \rightarrow {}^2\mathrm{He} \rightarrow d + e^+ + \nu_e \tag{4.24}$$

The intermediate nucleus ${}^2\mathrm{He}$ is highly unstable and it usually decays back to two protons rather than β-decaying into a deuteron. This means that reaction (4.24) occurs very infrequently. In the case of the CN cycle the carbon and nitrogen nuclei have relatively high electric charges and the reaction rate is lower than it would be if all the nuclei involved had low electric charge.

In fact deuterium, lithium, beryllium and boron will burn at lower temperatures than hydrogen because they can all burn without β-decays and without involving any particles with charges as high as carbon, nitrogen and oxygen. Bearing in mind that hydrogen is almost always the most abundant element, deuterium and the stable isotopes of lithium, beryllium and boron are destroyed by reactions involving protons. The actual reactions are as shown in (4.25):

$$\left. \begin{array}{l} d(p, \gamma){}^3\mathrm{He}, \\ {}^6\mathrm{Li}(p, {}^3\mathrm{He}){}^4\mathrm{He}, \\ {}^7\mathrm{Li}(p, \gamma){}^8\mathrm{Be} \rightarrow 2{}^4\mathrm{He}, \\ {}^9\mathrm{Be}(p, {}^4\mathrm{He}){}^6\mathrm{Li}(p, {}^3\mathrm{He}){}^4\mathrm{He}, \\ {}^{10}\mathrm{B}\,(p, {}^4\mathrm{He}){}^7\mathrm{Be}, \\ {}^{11}\mathrm{B}\,(p, \gamma)3{}^4\mathrm{He}, \end{array} \right\} \tag{4.25}$$

and the ^{7}Be produced in the fifth reaction is destroyed as in the PP chain. Although these reactions are expected to be the first that occur in stars and although they individually have a high yield of energy, they are not expected to play an important role in stellar structure (except that the first and third occur in the PP chain when the deuterium and lithium have themselves been built in the chain) because it is not believed that the elements concerned ever have high abundances. Thus I shall continue to regard the hydrogen burning reactions as the first significant ones to occur after a star is born.

It should be remarked that it is reactions involving deuterium rather than protons which are used in the hydrogen bomb and in experiments attempting to control thermonuclear reactions in the laboratory. Thus the suggested reaction chains for controlled thermonuclear reaction are:

$$\left.\begin{array}{l} d(d, p)^3H(d, n)^4He \\ (d, n)^3He(d, p)^4He. \end{array}\right\} \tag{4.26}$$

Recent estimates of the possibility of obtaining a useful energy release from controlled thermonuclear reactions suggest that it may be necessary for the initial fuel to be a mixture of deuterium and tritium (^{3}H), so that only one reaction in the above set is used. In that case, the tritium which is unstable but long lived would have to be produced first.

Helium burning reactions

If it is accepted that the burning of hydrogen will be the first nuclear process of importance in stellar evolution, there will come a time when there is no longer any hydrogen left in the central regions of the star. At this stage the central regions will still be hotter than the outer regions and thus energy will continue to flow outwards. As there is no longer any nuclear energy release, the energy flow can only come from the thermal energy. Any loss of thermal energy reduces the pressure of the stellar gas and the central regions are then compressed by the overlying layers. Provided the material remains an ideal classical gas, this compression leads to a rise in temperature and the rise in temperature continues until the next significant nuclear reactions occur; this may also be true even if the centre of the star does not remain an ideal gas.

The next significant reactions to occur are those involving ^{4}He, the product of hydrogen burning. It might be expected that the important reaction would involve the combination of two helium nuclei to produce either one or two other nuclei. Unfortunately this does not work as ^{8}Be is very unstable and decays to two helium nuclei again, as we have already seen in the reactions of the PP chain, while the other reactions involving two helium nuclei require energy rather than releasing energy. For some time it was unclear how further nuclear burning would proceed. Then it was realised that very rarely a third helium nucleus could be added to ^{8}Be before it decayed. Carbon is then formed by the chain:

$$\left.\begin{array}{l} ^4He + {}^4He \rightarrow {}^8Be, \\ ^8Be + {}^4He \rightarrow {}^{12}C + \gamma \end{array}\right\} \tag{4.27}$$

As with hydrogen burning reactions, this is an unusual reaction. In particular it is a resonant reaction, the addition of ^{4}He to ^{8}Be producing an excited state of ^{12}C. If the reaction were not resonant the addition of a further ^{4}He would convert ^{12}C to ^{16}O almost as soon as it was formed and the ^{12}C so important to life would not be present in anything like its observed quantity. The same would also be true if the conversion of carbon to oxygen went through a resonant reaction, which it does not.

Helium burning is effectively a three particle reaction so that the energy release per kg is proportional to the square of the density instead of being linearly proportional to density as in the case of hydrogen burning. The reaction rate is again very strongly dependent on temperature. In stars, helium burning occurs typically at temperatures of about 10^8K and near this temperature the rate of energy release is:

$$\varepsilon_{3\text{He}} \simeq \varepsilon_3 X_{\text{He}}^3 \rho^2 T^{40}, \tag{4.28}$$

where ε_3 is a constant and X_{He} is the fractional concentration of helium by mass.

Once helium is used up in the central regions of a star, further contraction and heating may occur and that may lead to additional nuclear reactions such as the burning of carbon. For the present I will not discuss these reactions, but will remark again that the majority of the possible energy release by nuclear fusion reactions has occurred by the time that hydrogen and helium have been burnt. The time that a star spends using up a particular nuclear fuel can be expected to be related to the amount of energy available in the reaction. There is however an additional factor that reduces the time that stars spend in later stages of stellar evolution. When the centre of a star is either very hot or very dense, reactions become possible which release neutrinos and anti-neutrinos in very large numbers. Because the neutrinos can escape essentially freely from a star, this adds substantially to the star's loss of energy and reduces significantly the time which a star can spend in that nuclear burning phase. This in turn reduces the probability that stars in such a stage of evolution will be observed. I shall mention the neutrino emission in late stages of stellar evolution further in Chapter 9.

Opacity

I now turn to a discussion of the opacity of stellar material. In Chapter 3 it has been stated that the flow of energy by conduction and radiation is essentially similar in nature and that, inside a star, the rate at which energy flows by these processes is determined by one quantity, the opacity κ. Thus we have (3.54):

$$\frac{\mathrm{d}T}{\mathrm{d}r} = -\frac{3\kappa L\rho}{16\pi acr^2 T^3},$$

which relates the rate of energy transport to the temperature gradient and the opacity. In the last chapter no formula was given for κ and a full discussion of how the formula is obtained is outside the scope of the present book. I will, however, discuss the principles underlying the determination of κ.

It has already been mentioned in Chapter 2 that, if matter and radiation are in equilibrium with one another at a temperature T, the radiation present is entirely described in terms of the Planck function $B_\nu(T)$ where

$$B_\nu(T) = \frac{2h\nu^3}{c^2} \frac{1}{\exp(h\nu/kT) - 1}. \tag{2.5}$$

$B_\nu(T)$ is the energy crossing an area of a square metre per second in a unit frequency range and unit solid angle. In thermodynamic equilibrium there is an equal flow or radiation in all directions. Inside a star conditions cannot be precisely those of thermodynamic equilibrium because in that case there would be no net flow of energy in the radial direction. However, the departures of the radiation intensity from the Planck function are very small indeed inside a star.

The opacity of stellar material is a measure of the resistance of the material to the passage of radiation; equivalently the radiative conductivity measures the ease with which energy flows. The probability that an individual photon will be absorbed depends on its frequency. Thus we can define the monochromatic mass absorption coefficient, κ_ν, of the material, which is such that radiation of intensity I_ν is changed by δI_ν in a distance δx, where

$$\delta I_\nu = -\kappa_\nu \rho I_\nu \delta x. \tag{4.29}$$

The dimensions of κ_ν are m^2 kg^{-1} and it is called a mass absorpion coefficient because of the inclusion of ρ in the definition (4.29). When considering the radiative conductivity of stellar material, it is reasonable to suppose that the effective conductivity will depend mainly on the conductivity in a frequency range in which the number of photons is a maximum. This mean that, in forming the average conductivity, the conductivity at any frequency should be multiplied by a quantity depending on the number of photons of that frequency which are present, before the average is calculated. As the opacity is essentially a reciprocal of the conductivity, this means that the reciprocal of the absorption coefficient, κ_ν, should be weighted with the number of photons present in forming the reciprocal of the opacity.

As mentioned above, energy only flows because the temperature is higher near the centre of the star. At any point the outward flowing radiation has been emitted at a slightly higher temperature and has a frequency distribution approximating a Planck function at that higher temperature. Similarly the inward flowing radiation has a frequency distribution corresponding to a Planck function at a slightly lower temperature. The net radiative conductivity is then obtained by multiplying the conductivity at any frequency by the difference between these two Planck functions, by integrating over frequency and by suitably normalising the answer. In terms of opacity rather than conductivity

$$\frac{1}{\kappa} = \int_0^\infty \frac{1}{\kappa_\nu} \frac{dB_\nu}{dT} d\nu \bigg/ \int_0^\infty \frac{dB_\nu}{dT} d\nu. \tag{4.30}$$

The effective opacity is calculated from this formula once the absorption coef-

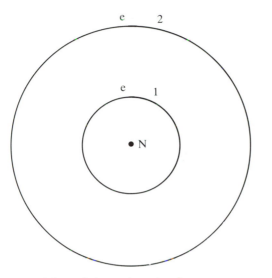

Fig. 40. Bound–bound absorption. An electron e can move from orbit 1 to orbit 2 about a nucleus N, with absorption of a photon.

ficient at all frequencies is known. A somewhat more detailed derivation of this equation and of the equation of radiative transfer is given in Appendix 2.

The sources of opacity

The discussion of stellar opacity now involves considering all of the microscopic processes which contribute to the absorption of radiation of frequency ν. It is impossible to discuss these in detail in this book, but an account can be given of the basic types of process involved. There are four of these:

    (i) bound–bound absorption,
    (ii) bound–free absorption,
    (iii) free–free absorption.
    (iv) scattering.

These processes are described below. The first three are called true absorption processes because they involve the disappearance of a photon, while in the fourth case only the direction of motion of the photon is altered. In addition there is the contribution to the effective opacity from thermal conduction. This will not be discussed explicitly below, but it is included in the numerical calculations which are mentioned.

Bound–bound absorption

In this case an electron is moved from one bound orbit in an atom or ion to an orbit of higher energy with the absorption of a photon (fig. 40). Thus, if the

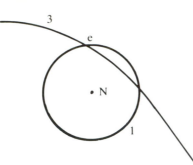

Fig. 41.   Bound–free absorption. The absorption of a photon can move an electron from the bound orbit 1 to the free orbit 3.

energy of the electron in the two orbits is E_1 and E_2 respectively, a photon of frequency ν_{BB}. can produce the transition if

$$E_2 - E_1 = h\nu_{BB}. \tag{4.31}$$

These bound–bound processes, which are responsible for the characteristic spectral lines of different elements and which lead to the occurrence of spectral lines in the visible radiation from stars, are not very important in the deep interiors for two reasons. As all of the atoms are highly ionised, only a small minority of electrons are in bound states. In addition the majority of photons have energy somewhere near that corresponding to the maximum of the Planck function (2.5), which occurs at frequency ν_{max} satisfying:

$$h\nu_{max}/kT = 2.82. \tag{4.32}$$

For conditions in deep stellar interiors the energy $h\nu_{max}$ ($=2.82kT$) is in many cases greater than the separation in energy between atomic bound states and the photons are more likely to cause a bound–free absorption, which I now describe.

Bound–free absorption

This involves an electron in a bound state around a nucleus being moved into a free hyperbolic orbit by the absorption of a photon (fig. 41) A photon of frequency ν_{BF} can be absorbed and convert a bound electron of energy E_1 into a free electron of energy E_3 provided that

$$E_3 - E_1 = h\nu_{BF}. \tag{4.33}$$

In this case, as in the case of bound–bound absorption, the importance of the process is reduced because of the scarcity of bound electrons. However, provided the photon has sufficient energy to remove the electron from the atom, any value of energy can lead to a bound–free process.

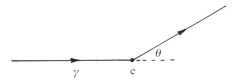

Fig. 42.   The scattering of a photon γ by an electron e.

Free–free absorption

In this case the electron is initially in a free state with energy E_3 and it absorbs a photon of frequency ν_{FF} and moves to a state with energy E_4, where

$$E_4 - E_3 = h\nu_{FF}. \tag{4.34}$$

There is no restriction on the energy of a photon which can induce a free–free transition but, in both free–free and bound–free absorption, it is found that low energy photons are more likely to be absorbed than high energy photons.

Later in this chapter I shall explain that in very dense stars the energy states which electrons can occupy are affected by the quantum mechanical Pauli exclusion principle which says that no more than one electron can occupy any quantum state. In a dense gas all of the low energy states are occupied by electrons. In that case the energy which a photon must possess to cause either a bound–free or a free–free transition must be sufficient to put the electron into a state which is not already occupied.

Scattering

Finally it is possible for a photon to be scattered by an electron or an atom. On a classical picture, this process can be idealized as a collision between two particles which bounce off one another. If the energy of the photon satisfies

$$h\nu \ll mc^2, \tag{4.35}$$

where m is the mass of the particle doing the scattering, the particle is scarcely moved by the collision. In this case the photon can be imagined to be bounced off a stationary particle (fig. 42). The inequality is valid throughout most stars, but is violated in stars with extremely high internal temperatures. Although this process does not lead to the true absorption of radiation it does slow down the rate at which energy escapes from a star because it continually changes the direction of the photons.

The calculation of stellar opacity is a very complicated process as all atoms and ions must be considered. Because (4.30) for κ involves what is known as a *harmonic mean* of κ_ν rather than a direct mean, I cannot calculate a mean absorption coefficient for each chemical element independently and then add the results together to form κ. Instead all of the contributions to κ_ν must be added together before the mean is calculated. This means that each time I wish to consider a star of a different chemical composition, (4.30) must be calculated afresh. Another property of (4.30) for κ is that a sensible value of κ can only be obtained if we have an estimate of the monochromatic absorption coefficient at all frequencies. If, in (4.30), κ_ν is put equal to zero in any frequency band, the resulting value of κ is zero, which suggests that radiation escapes freely. This is clearly not true, if κ_ν is non-zero for most frequencies. However there must be some absorption at all frequencies if the intensity is to remain close to the Planck function.

Numerical values for opacity

There is a slight subtlety in (4.30) for the opacity. Associated with any absorption process, but not scattering, there is another process known as stimulated emission. The absorption of photons is followed by the emission of photons of the same frequency moving in the same direction as the incident radiation. This reduces the effective absorption coefficient which must be included in the opacity. In conditions close to thermodynamic equilibrium κ_ν is replaced by $\kappa_\nu(1 - \exp(-h\nu/kT))$. To this must be added the mass scattering coefficient σ_ν.

Further details of the calculation of opacities cannot be given here, but the results of a detailed calculation are shown in fig. 43. For material of one particular chemical composition, the opacity is shown as a function of temperature and density. It can be seen that the opacity is low at both very high and low temperatures. At high temperatures most of the photons have high energy and, as has been mentioned previously, they are absorbed less easily than lower energy photons. At low temperatures most atoms are not ionised and there are few electrons available to scatter radiation or to take part in free–free absorption processes, while most photons have insufficient energy to ionise atoms. The opacity has a maximum at intermediate temperatures where bound–free and free–free absorption are very important. Very recent calculations have shown that opacities at low temperatures ($\lesssim 10^6$K) have been underestimated in earlier calculations such as those illustrated in fig. 43. For some stars at some stages of evolution these corrections to the opacity are very important.

From fig. 43, one result of immediate interest can be obtained. In the Sun the central density is about 10^5 kg m^{-3} and the opacity is about 10^{-1} m^2 kg^{-1}, so that $\kappa\rho \simeq 10^4$ m^{-1}. From (4.29), this means that a typical photon travelling from the centre of the Sun is absorbed or scattered when it has travelled about 10^{-4} m. Further out in the Sun, when the density is about 10^3 kg m^{-3}, the opacity is 10 m^2 kg^{-1}, so that once again the mean free path of radiation is about 10^{-4} m. The central temperature of the Sun is believed to be about 1.5×10^7K so that the mean temperature gradient in the Sun is about 2×10^{-2}K m^{-1}. This means that a typical temperature difference between the point at which a photon is emitted and the point at which it is absorbed is 2×10^{-6}K. This is remarkably small and this is the reason why the intensity function I_ν is able to remain so very close to the Planck function $B_\nu(T)$ in stellar interiors. The departure from true thermodynamic equilibrium is very slight.

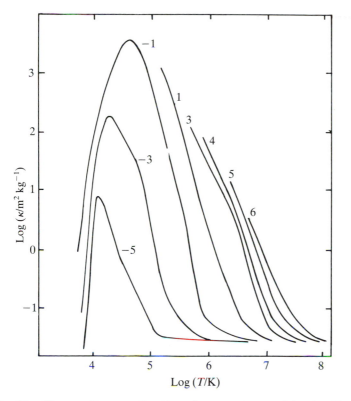

Fig. 43. The opacity κ as a function of temperature and density. Each curve represents a different value of the density and is labelled by log (ρ/kgm^{-3}).

Approximate form for opacity

Approximate analytical expressions for the opacity in particular ranges of temperature and density can be read off from the curves of fig 43. At high temperatures, there is a range of density for which the opacity is scarcely dependent on either temperature or density (although there is a dependence on chemical composition which cannot be shown in fig. 43), so that to a first approximation I may write:

$$\kappa = \kappa_1, \tag{4.36}$$

where κ_1 is a constant for stars of a given chemical composition. At high temperatures the main source of opacity is the scattering of radiation by free electrons (Compton scattering) and, if no other processes are important, the opacity has precisely the form (4.36).

At lower temperatures the processes of bound–free and free–free absorption become important and there is a range of temperature in which the opacity increases with increasing density and with decreasing temperature. A reasonable analytical approximation to the opacity has the form:

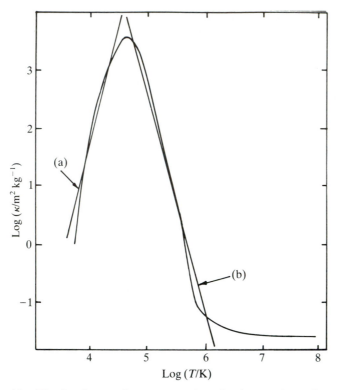

Fig. 44.   The fit of approximate expressions for the opacity to the true expression. The curve is for a density of $10^{-1}\mathrm{kg m}^{-3}$. The straight line (a) is the approximation (4.38), the straight line (b) is the approximation (4.37) and the opacity approaches a constant value, approximation (4.36), at high temperature.

$$\kappa = \kappa_2 \rho / T^{3.5}, \tag{4.37}$$

where κ_2 is again a constant for stars of given chemical composition. At even lower values of the temperature, the opacity decreases with decreasing temperature and an approximate analytical form for the opacity in that regime is:

$$\kappa = \kappa_3 \rho^{1/2} T^4, \tag{4.38}$$

where κ_3 is another constant. This region of the opacity curve is not important in stars whose surface temperature is high enough for the abundant elements, hydrogen and helium, to be significantly ionised even at the stellar surface. Figure 44 shows how these approximate expressions (4.36)–(4.38) fit to one of the calculated curves of fig. 43.

When calculations were first made of the structure and evolution of the stars, simple power law approximations to the law of energy release, such as:

$$\varepsilon = \varepsilon_0 \rho T^\eta, \tag{4.23}$$

and to the law of opacity:

$$\kappa = \kappa_0 \rho^{\lambda-1} / T^{\nu-3} \tag{4.39}$$

(where the choice of the exponents $\lambda-1$, $\nu-3$ is made to give a simple form in (3.54), and ν must not be confused with the frequency) were extremely important. Their use enabled progress to be made in the subject without the use of electronic computers, which had not yet been developed. Indeed, as we shall see in the next chapter, quite useful results could be obtained without the aid of any type of calculating machine. Today the development of powerful electronic computers enables much more detailed information about the laws of opacity and energy generation to be used in the computations. Indeed it is now possible to solve the equations of stellar structure quickly on desk top computers.

The equation of state of stellar material

The third quantity whose behaviour must be considered is the pressure. I have already stated that stellar material is gaseous and that in many cases it behaves like an ideal classical gas. If it is an ideal gas, the gas pressure is:

$$P_{gas} = nkT, \qquad (3.25)$$

where n is the number of particles per cubic metre and k is Boltzmann's constant. To obtain this in the form

$$P_{gas} = P(\rho, T, \text{composition}),$$

I require an expression for n in terms of ρ, T and composition. Astronomers usually write (3.25) in the alternative form by introducing:

$$\mu \equiv \rho/nm_H, \qquad (4.40)$$

where m_H is the mass of the hydrogen atom (1.67×10^{-27} kg), so that μ is the mean mass of the particles in the gas in terms of the mass of the hydrogen atom. μ is called the *mean molecular weight* of the stellar material. Then introducing the gas constant:

$$\mathcal{R} = k/m_H, \qquad (4.41)$$

where $\mathcal{R} = 8.26 \times 10^3$ J K^{-1} kg^{-1},† (3.25) takes the form:

$$P_{gas} = \mathcal{R}\rho T/\mu, \qquad (4.42)$$

which is the form I shall use subsequently in this book.

Mean molecular weight of ionised gas

I now require an expression for μ as a function of ρ, T and chemical composition. The calculation of μ for completely general values of ρ and T is very complicated because to obtain a value of n the fractional ionisation of all the elements has to be computed. Fortunately, throughout most of the interior of

---

† Note that $\mathcal{R}$ differs from R by a factor of approximately 10^3. They are essentially equal in c.g.s. units, but differ in SI units because, while the unit of mass is different, the mass of a mole is not.

stars, the expression can be simplified for two reasons. In the first place all of the elements are highly ionised and, secondly, hydrogen and helium are very much more abundant than all of the other elements and they are certainly fully ionised in stellar interiors. This means that only a very small error is made in the value of μ if it is assumed that all of the material is fully ionised. Near the stellar surface this approximation becomes inadequate and a more careful discussion of the value of μ is required. This is important for detailed calculations of stellar structure, but for approximate discussion in Chapter 5 only the value of μ for a fully ionised gas will be required.

If the material is assumed to be fully ionised, the calculation of μ proceeds as follows. Suppose that the fractional abundances by mass of hydrogen, helium and all elements other than hydrogen and helium are called X, Y and Z (instead of X_H, X_{He}, etc. as defined above) so that

$$X + Y + Z = 1. \tag{4.43}$$

This means that in a cubic metre of material of density ρ there is a mass $X\rho$ of hydrogen, a mass $Y\rho$ of helium and a mass $Z\rho$ of heavier elements.

In a cubic metre there are therefore $X\rho/m_H$ hydrogen atoms. Each ionised hydrogen atom consists of two particles, a proton and an electron, so that the hydrogen provides $2X\rho/m_H$ particles per cubic metre. In a similar way, as the mass of a helium atom is $4m_H$, there are $Y\rho/4m_H$ helium atoms per cubic metre, Each ionised helium atom consists of three particles and the helium provides $3Y\rho/4m_H$ particles per cubic metre. In principle each heavy element should be considered separately. However, for these heavier elements, the number of electrons is always about half the atomic mass number in units of m_H. Thus to a reasonable approximation when they are fully ionised they supply about one particle for every $2m_H$; the number is somewhat larger for atoms of low atomic weight and lower for the heavier atoms. To a first approximation the number of particles provided by the heavy elements is $Z\rho/2m_H$.

The total number of particles per cubic metre can then be written:

$$n = (\rho/m_H)[2X + 3Y/4 + Z/2]. \tag{4.44}$$

Using (4.43), expression (4.44) can be rewritten:

$$n = (\rho/4m_H)[6X + Y + 2]. \tag{4.45}$$

This can be combined with (4.40) to give:

$$\mu = 4/(6X + Y + 2), \tag{4.46}$$

which is a good approximation for μ except in the cool outer regions of a star. In many cases the fractional abundance of the heavy elements is so small that Z can be neglected in (4.46) and Y replaced by $1 - X$ to give:

$$\mu = 4/(3 + 5X), \tag{4.47}$$

and this expression for μ is used in the next chapter. The inaccuracy introduced in using such an approximate expression for μ is much less than inevitable inaccuracies in quantities such as κ and in the mixing length theory of convection.

Departures from ideal classical gas law

Expressions (4.46) and (4.47) give a good approximation to the mean molecular weight of a fully ionised gas and, while the gas remains ideal classical, its pressure can then be found from (4.42). I have mentioned earlier that as a star evolves it tends to contract and heat up and that this process continues while the stellar gas remains ideal classical. Is there any reason why it should not remain perfect? We can expect to find departures from the perfect gas law at high enough densities when the particles in the gas are packed close together, as in the case of the well-known van der Waals' forces.

In fact, in the ionised gas in stellar interiors, the first deviation occurs because electrons have to obey Pauli's exclusion principle. In its most familiar form this principle states that no more than one electron can occupy any one bound energy state in an atom.† The principle plays an important role in the arrangement of electrons in atoms and in the explanation of the periodic table of the elements. The Pauli exclusion principle also places a restriction on the relative position and momentum of two free electrons which are not attached to atoms, and to free protons and neutrons. Heisenberg's principle of uncertainty says that the position and momentum of any one particle cannot be measured simultaneously with complete accuracy. There is an uncertainty δx in any position coordinate and δp in the corresponding momentum coordinate which must satisfy

$$\delta x \delta p \geq h/4\pi. \tag{4.6}$$

The particle is, so to speak, surrounded by an uncertainty volume in the six-dimensional space which has three position coordinates and three momentum coordinates. In fact a volume h^3 in this space corresponds to one quantum state. This means that, if particles are closely packed together in position space, they may be forced to have a higher momentum than is predicted by the kinetic theory of gases for a classical gas. As a result of this a gas at a given temperature and density has a higher total internal energy and a higher pressure than is predicted by the classical gas law. A gas in which the Pauli exclusion principle is important is called a *degenerate gas*. Because at a given temperature ions have a higher momentum than electrons, the ions are less likely to be in danger of violating the Pauli exclusion principle. In stars electrons may form a degenerate gas, but the ions can almost always be treated as a classical gas. However, neutrons may be degenerate in a neutron star. The derivation of the formula for the pressure of a degenerate gas is sketched in Appendix 3 and the result of the calculation is as follows.

---

† In fact if the property known as electron spin is allowed for, two electrons with oppositely directed spin can be in any one state.

At a high enough density, the momentum of a particle is essentially determined by the Pauli exclusion principle rather than by the temperature of the gas. This means that the pressure and internal energy of the gas become essentially independent of temperature. The precise form of the pressure depends on whether the highest momentum possessed by a particle is greater than, or less than $m_e c$, where m_e is the electron mass and c the velocity of light. The maximum possible velocity of an electron is, of course, c but according to the special theory of relativity the momentum p, can exceed $m_e c$ as it is given by the formula:

$$p = m_e v / (1 - v^2/c^2)^{1/2} \tag{4.48}$$

which shows that p increases without limit as v approaches c.

If the maximum electron momentum, p_0, satisfies $p_0 \ll m_e c$, the pressure can be shown to be:

$$P_{gas} \simeq K_1 \rho^{5/3}, \tag{4.49}$$

where

$$K_1 = \frac{h^2}{20 m_e} \left(\frac{3}{\pi}\right)^{2/3} \left(\frac{1 + X}{2 m_H}\right)^{5/3}. \tag{4.50}$$

As before, X is the mass fraction in the form of hydrogen and the gas has been assumed to be fully ionised. If the momentum p_0 satisfies $p_0 \gg m_e c$, the pressure is given by:

$$P_{gas} \simeq K_2 \rho^{4/3} \tag{4.51}$$

where

$$K_2 = \frac{hc}{8} \left(\frac{3}{\pi}\right)^{1/3} \left(\frac{1 + X}{2 m_H}\right)^{4/3}. \tag{4.52}$$

Of course there must be gradual progression between the two formulae (4.49) and (4.51) for intermediate values of $p_0/m_e c$, which is itself determined by the relation:

$$\frac{p_0}{m_e c} = \left(\frac{3 h^3 \rho (1 + X)}{16 \pi m_H m_e^3 c^3}\right)^{1/3} \tag{4.53}$$

Similarly there is not a sharp transition between the ideal classical gas formula (4.42) and the formulae (4.49) and (4.51); there is a region of temperature and density in which some intermediate and much more complicated formula must be used. When the pressure of a gas is given by (4.49) it is said to be *non-relativistically degenerate*; when it is given by (4.51) it is said to be *relativistically degenerate*.

In addition to the pressure of the particles, we must consider the radiation pressure. The expression

$$P_{rad} = \tfrac{1}{3} a T^4 \tag{3.27}$$

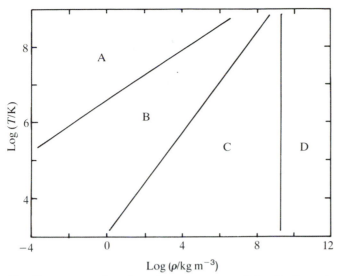

Fig. 45.   Pressure as a function of temperature and density. In region A radiation pressure is larger than gas pressure, in region B the material behaves as an ideal classical gas, in region C the non-relativistic degenerate law holds and in region D the relativistic degenerate approximation is valid.

is a valid expression for the radiation pressure provided that the radiation is essentially distributed according to the Planck formula. Although this is usually true, there are conditions in stellar interiors when the distribution of radiation with frequency departs seriously from the Planck Law. Under these conditions (3.27) may not be a good approximation to the radiation pressure. It appears, however, that almost always, when the formula (3.27) for radiation pressure is not a good approximation, the gas pressure is very much larger than the radiation pressure and the precise value of the radiation pressure is unimportant. Thus I shall assume that (3.27) is always the correct expression for the radiation pressure.

For a fully ionised gas of a given chemical composition, it is possible to divide the log ρ–log T plane into regions in which radiation pressure is more important than gas pressure and in which gas pressure is more important than radiation pressure. The latter region can be subdivided into regions when the simple ideal gas law (4.42) holds, where the non-relativistic degeneracy formula (4.49) is a good approximation and where the relativistic degeneracy formula (4.51) holds. These regions are shown in fig. 45.

One of the main reasons for interest in the fact that the stellar gas becomes degenerate at high enough densities arises because of the consequences this has for the Virial Theorem:

$$3 \int (P/\rho) \, \mathrm{d}M + \Omega = 0. \tag{3.24}$$

I have previously used this equation to show that any contraction of a star made of an ideal classical gas leads to its heating up. This arises because, as a star contracts Ω must become more negative and thus the average value of P/ρ must increase.

For a classical gas of constant mean molecular weight μ this implies that T must increase. Once the stellar material is ionised any change in μ is likely to be an increase due to nuclear fusion reactions and this makes it even more definite that the temperature will increase. Once the stellar gas becomes degenerate, it is possible for P/ρ to increase at the same time as T decreases, because to a first approximation P/ρ given by formulae (4.49) and (4.51) increases as ρ increases and is independent of T. This has the consequence that once the central regions of a star become degenerate they may reach a maximum temperature and then begin to cool down. It this happens it no further nuclear fusion reactions will occur in the star and its luminosity will decrease as it contracts until it finally becomes invisible. I have not shown that this will definitely happen, but it now appears that there is a chance that the central temperatures of stars will not rise irrevocably as they evolve. We have already seen in Chapter 2 that the white dwarfs are faint stars of very high density and it is believed that they are cooling degenerate stars and that they represent the final stages of stellar evolution. They will be discussed in Chapter 10. I shall also explain in the next chapter that stars of sufficiently low mass never become hot enough for fusion reactions converting hydrogen to helium.

Summary of Chapter 4

In this chapter I have discussed how pressure, opacity and rate of energy generation depend on temperature, density and chemical composition.

Energy can be released by nuclear fusion reactions building light elements up to nuclei around iron, but most of the energy is released by the conversion of hydrogen into helium. Because positively charged nuclei repel one another, they must have high velocities to approach close enough for short-range nuclear forces to cause a nuclear reaction. Thus nuclear reactions in significant numbers only occur when stellar temperatures are high. The rates of nuclear reactions depend on a high power of the temperature and reactions involving light elements occur at lower temperatures than those involving heavier elements. As a star's central temperature rises while it remains an ideal classical gas, a succession of nuclear reactions can be expected to occur, with the most important hydrogen burning reactions being the first.

The opacity of stellar material is determined by all the processes which scatter and absorb photons. These include the scattering of radiation by electrons, and the absorption of photons by an atom which causes either a bound electron to move to another bound orbit, or to escape, or an electron to move from one free orbit to another of higher energy. The calculation of opacity requires values for the rates at which many of these processes occur.

Stellar material often behaves as an ideal classical gas and the value of its pressure can be calculated from Boyle's law. When its density becomes high, the electrons may become so close together that the Pauli exclusion principle restricts their possible momenta. When this is so, the material ceases to be an ideal classical gas and becomes a degenerate gas. In a highly degenerate gas, the pressure depends only on density and chemical composition and it is then possible for a star to cool down as it loses energy.

Completely accurate expressions for the opacity, rate of energy generation and pressure are extremely complicated, but in some ranges of temperature and density it is possible to find simple mathematical expressions which are good approximations to the physical quantities. These approximate expressions are used in Chapter 5 to obtain qualitative information about the structure of the stars.

5

The structure of main sequence stars and pre-main-sequence evolution

Introduction

In Chapter 2 we have seen that the majority of stars are main sequence stars and I have already suggested that this could mean either that most stars are main sequence stars for all of their lives or that all stars spend a considerable fraction of their life in the main sequence state. It is now believed that the latter is true and that the main sequence phase is one in which stars are obtaining their energy from the conversion of hydrogen into helium which, as we have seen in the last chapter, releases 83% of the maximum energy which can be obtained from nuclear fusion reactions. It is also believed that main sequence stars are chemically homogeneous, which means that there are no significant variations of chemical composition from place to place within the stars. As hydrogen burning is the first important nuclear reaction to occur as the central temperature of a star rises, stars should be chemically homogeneous when they reach the main sequence, provided the same was true of the interstellar cloud out of which they were formed. In this chapter I consider the structure of such stars. In the last two chapters I have discussed all of the relevant equations and have concluded that in general their solution can only be obtained with the use of a large computer. However, it is possible to obtain some general properties of chemically homogeneous stars without solving the equations and, in particular, a qualitative understanding of the existence and position of the main sequence in the HR diagram (fig. 46) and of the mass–luminosity relation can be obtained.

In the main part of this chapter I shall discuss the properties of main sequence stars without asking how the stars reach the main sequence and whether the main sequence properties depend on their previous life history. In fact, it is believed that for most stars main sequence stellar structure is almost independent of previous life history. Provided no significant nuclear reactions occur before the main sequence is reached, the stars on the main sequence have the chemical

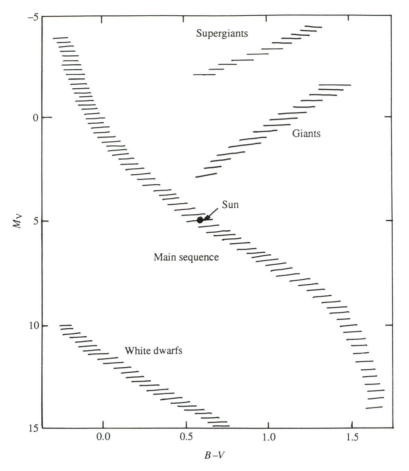

Fig. 46.   The Hertzsprung–Russell diagram for nearby stars.

composition with which they were formed from the interstellar gas. As the properties of stars change very slowly in the main sequence phase, they can be studied by solving a set of equations in which time dependence does not explicitly enter and which therefore make no reference to the previous history of the star. It is fortunate that main sequence stellar structure is almost independent of previous life history because even now the process of star formation is not very well understood. It has proved possible to study post-main-sequence evolution at a time when very little was known about pre-main-sequence evolution. Some remarks about star formation and pre-main-sequence evolution will be made at the end of this chapter.

The structure of chemically homogeneous stars

In this chapter I first study the structure of stars which have the same chemical composition, but which differ in mass. Subsequently I shall ask how the

, properties of stars depend on their chemical composition. The equations of stellar structure, which I have discussed in Chapters 3 and 4, are too complicated for me to hope to find an exact analytical solution of them and they must be solved by use of a computer. However, provided certain approximations are made, it is possible to discover how properties of a star such as luminosity, radius and effective temperature change, if the mass of the star under consideration is altered, without solving the equations completely. Although this procedure is not a substitute for a complete solution of the problem, the results obtained provide a very useful check on more detailed computations.

The approximations are concerned with the three equations:

$$P = P(\rho, T, \text{composition}), \tag{3.70}$$

$$\kappa = \kappa(\rho, T, \text{composition}) \tag{3.71}$$

and

$$\varepsilon = \varepsilon(\rho, T, \text{composition}), \tag{3.72}$$

which relate the pressure, opacity and rate of energy generation to the density, temperature and chemical composition of the stellar material. I suppose that radiation pressure can be neglected and that the stellar material behaves as an ideal classical gas so that an expression for the pressure is the perfect gas law:

$$P = \mathcal{R}\rho T/\mu. \tag{5.1}$$

In Chapter 4 I showed that, for appropriate ranges of temperature and density, the laws of opacity and energy generation can be approximately represented by power laws:

$$\kappa = \kappa_0 \rho^{\lambda-1}/T^{\nu-3} \tag{4.39}$$

and

$$\varepsilon = \varepsilon_0 \rho T^\eta, \tag{4.23}$$

where λ, ν and η are constants and κ_0 and ε_0 are constants for a given chemical composition. In this chapter I first assume that (5.1), (4.39) and (4.23) are accurately true. I also suppose in the first instance that no energy is carried by convection and that, in addition, I may use the simplest boundary conditions discussed in Chapter 3,

$$r = 0, \quad L = 0 \quad \text{at} \quad M = 0 \tag{3.79}$$

and

$$\rho = 0, \quad T = 0 \quad \text{at} \quad M = M_s, \tag{3.80}$$

where M_s is the total mass of the star being considered. Although it is unlikely that a single approximation for the opacity of the form (4.39) will hold throughout the interior of a star, the results should still provide some approximation to the properties of real stars.

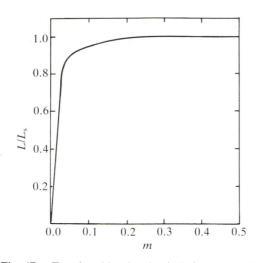

Fig. 47.   Fractional luminosity (L/L_s) as a function of fractional mass.

The equations of stellar structure

$$\frac{\mathrm{d}P}{\mathrm{d}M} = -\frac{GM}{4\pi r^4},$$

(3.75)

$$\frac{\mathrm{d}r}{\mathrm{d}M} = \frac{1}{4\pi r^2 \rho},$$

(3.76)

$$\frac{\mathrm{d}L}{\mathrm{d}M} = \varepsilon$$

(3.77)

and

$$\frac{\mathrm{d}T}{\mathrm{d}M} = -\frac{3\kappa L}{64\pi^2 acr^4 T^3}$$

(3.78)

must now be solved with the boundary conditions (3.79) and (3.80) and the supplementary relations (5.1), (4.39) and (4.23).

For a sequence of stars of the same homogeneous chemical composition, for which κ_0, ε_0 and μ are the same for all of the stars, I can now show that the properties of a star of any mass can be deduced once the properties of a star of any one mass are known. What I shall in fact show is that the way in which any physical quantity such as luminosity varies from the centre of the star to the surface is the same for stars of all masses and only the absolute value of the luminosity varies from star to star. This is illustrated schematically in fig. 47 where the ratio of luminosity to surface luminosity (L/L_s) is plotted against the *fractional mass*:

$$m \equiv M/M_s.$$

(5.2)

I shall show that the curve shown in fig. 47 is the same for all stars with the same laws of opacity and energy generation, but that the value of L_s depends on M_s and that it is proportional to some power of M_s, which depends on the values of λ, v, and η in (4.39) and (4.23). The same is also true of other quantities such as radius (r_s), effective temperature (T_e) and central temperature (T_c). The equations of stellar structure need only be solved once and the properties of stars of all masses can then be deduced.

My statement above about the luminosity is equivalent to saying that the luminosity of any point inside a star depends on some power of M_s but otherwise only on the fractional mass m. Mathematically this (and similar statements for the other physical quantities) implies that

$$
\left.
\begin{aligned}
r &= M_s^{a_1}\bar{r}(m), \\
\rho &= M_s^{a_2}\bar{\rho}(m), \\
L &= M_s^{a_3}\bar{L}(m), \\
T &= M_s^{a_4}\bar{T}(m)
\end{aligned}
\right\}
\tag{5.3}
$$

and

$$
P = M_s^{a_5}\bar{P}(m),
$$

where a_1, a_2, a_3, a_4 and a_5 are constants and where (as indicated) $\bar{r}$, $\bar{\rho}$, $\bar{L}$, $\bar{T}$ and $\bar{P}$ depend only on the fractional mass m. I now verify that expressions of the form of (5.3) do satisfy the equations of stellar structure provided that values of the constants a_1 to a_5 are chosen correctly.

I consider each equation in turn. Equation (3.75) can be written:

$$
M_s^{a_5-1}\,d\bar{P}/dm = -GM_s^{1-4a_1}m/4\pi\bar{r}^4.
$$

If this equation is to be true for all values of M_s, the powers of M_s entering on the two sides of the equation must be the same. Thus:

$$
4a_1 + a_5 = 2,
\tag{5.4}
$$

and the equation becomes:

$$
d\bar{P}/dm = -Gm/4\pi\bar{r}^4.
\tag{5.5}
$$

Similarly (3.76) can be written:

$$
M_s^{a_1-1}\,d\bar{r}/dm = 1/4\pi M_s^{2a_1+a_2}\bar{r}^2\bar{\rho},
$$

which will reduce to an equation independent of M_s if

$$
3a_1 + a_2 = 1,
\tag{5.6}
$$

and which then has the form:

$$
d\bar{r}/dm = 1/4\pi\bar{r}^2\bar{\rho}.
\tag{5.7}
$$

Before separating (3.77) into an algebraic equation and a differential equation, the expression (4.23) for ε can be substituted on the right-hand side. Then the equation can be written:

$$M_s^{a_3-1}\, d\bar{L}/dm = \varepsilon_0 M_s^{a_2+\eta a_4}\bar{\rho}\bar{T}^{\eta}.$$

This is independent of M_s if

$$a_3 = 1 + a_2 + \eta a_4, \tag{5.8}$$

and then

$$d\bar{L}/dm = \varepsilon_0\bar{\rho}\bar{T}^{\eta}. \tag{5.9}$$

If (4.39) for κ is substituted into (3.78), this can be written:

$$M_s^{a_4-1}\, d\bar{T}/dm = -3\kappa_0 M_s^{[a_3+(\lambda-1)a_2-4a_1-\nu a_4]}\bar{\rho}^{(\lambda-1)}\bar{L}/64\pi^2 ac\bar{r}^4\bar{T}^{\nu}.$$

This is independent of M_s if

$$4a_1 + (\nu + 1)a_4 = (\lambda - 1)a_2 + a_3 + 1, \tag{5.10}$$

and then

$$d\bar{T}/dm = -3\kappa_0\bar{\rho}^{(\lambda-1)}\bar{L}/64\pi^2 ac\bar{r}^4\bar{T}^{\nu}. \tag{5.11}$$

Finally (5.1) gives:

$$a_5 = a_2 + a_4 \tag{5.12}$$

and

$$\bar{P} = \Re\bar{\rho}\bar{T}/\mu. \tag{5.13}$$

I now have five algebraic equations (5.4), (5.6), (5.8), (5.10) and (5.12) for the constants $a_1 \ldots a_5$. These are *inhomogeneous algebraic equations* which means that some of them contain terms which are independent of the as. Such a system of inhomogeneous equations will have consistent solutions provided that the determinant of the coefficients of $a_1 \ldots a_5$ does *not* vanish. The condition for the determinant to vanish is that

$$\nu - 3\lambda = \eta + 3, \tag{5.14}$$

and this is not true for any of the approximate laws of opacity and energy generation which we have considered in Chapter 4; η is usually large and positive while $\nu - 3\lambda$ is close to zero. Thus it appears that (5.4), (5.6), (5.8), (5.10) and (5.12) can be solved to give unique values for the constants $a_1 \ldots a_5$. Because the general solution is complicated it will not be written down, but solutions will shortly be given for special values of λ, ν and η.

To obtain the details of the structure of a star of any given mass the differential equations (5.5), (5.7), (5.9) and (5.11) and the equation (5.13) must now be solved to find $\bar{r}$, $\bar{\rho}$, $\bar{L}$, $\bar{T}$ and $\bar{P}$ in terms of m. The centre and surface of the star are $m = 0$ and $m = 1$ respectively and the boundary conditions are

$$\bar{r} = \bar{L} = 0 \text{ at } m = 0 \tag{5.15}$$

and

$$\bar{\rho} = \bar{T} = 0 \text{ at } m = 1. \tag{5.16}$$

This set of equations can now be solved on a computer and after the solution has been obtained the quantities $\bar{r}$, $\bar{\rho}$, etc. can be converted into r, ρ, etc. for a star of any given mass M_s by using the relations (5.3) and the values of the constants $a_1 \ldots a_5$ previously found. As mentioned earlier, the equations have only to be solved once and the properties of stars of all masses can then be obtained. Such a set of models of stars in which the dependence of the physical quantities on fractional mass m is independent of the total mass of the star is known as a *homologous sequence of stellar models*.

Mass–luminosity and luminosity–effective temperature relations

For such homologous stellar models there is clearly a mass–luminosity relation and also a simple relation between luminosity and effective temperature such as that which characterizes the main sequence in the HR diagram. Thus I have shown that at any point inside such a star:

$$L = M_s^{a_3}\bar{L}(m).$$

At the surface of the star ($m = 1$) this equation becomes:

$$L_s = M_s^{a_3}\bar{L}(1).$$

Since $\bar{L}(1)$ is the same for all stars of the same chemical composition, the luminosity should be proportional to the a_3th power of the mass. Values of a_3 will be considered shortly and they will be found to give a mass–luminosity relation similar to that found observationally for main sequence stars.

In addition

$$r_s = M_s^{a_1}\bar{r}(1), \tag{5.18}$$

while

$$L_s = \pi a c r_s T_e^4. \tag{2.7}$$

Combining (5.17), (5.18) and (2.7) it can be seen that

$$T_e = M_s^{(a_3-2a_1)/4}[\bar{L}(1)/\pi a c \bar{r}^2(1)]^{1/4}. \tag{5.19}$$

Equations (5.17) and (5.19) then show that for the homologous sequence of stars:

$$L_s \propto T_e^{4a_3/(a_3-2a_1)}. \tag{5.20}$$

This shows that the stars lie on a straight line in the theoretical HR diagram (plot of log L_s against log T_e) and this might be identified with the main sequence. Note that the dependence of T_e on M_s shown in (5.9) is not in general the same as the dependence of T on mass in (5.3).

Solutions for particular laws of opacity and energy generation

Two things must now be considered. Do the values of λ, ν and η, which have been suggested in Chapter 4, lead to a theoretical mass–luminosity relation and main sequence which are in reasonable agreement with the observations, and to what extent are the various assumptions which led to the homologous sequence of models valid? I first study the homologous solutions in more detail and then discuss their limitations.

In Chapter 4 two particular approximations to the opacity law which have been mentioned are (equations (4.36) and (4.37)):

$$\kappa = \kappa_1 \qquad \lambda = 1, \qquad \nu = 3 \tag{5.21}$$

and

$$\kappa = \kappa_2 \rho / T^{3.5}, \qquad \lambda = 2, \qquad \nu = 6.5. \tag{5.22}$$

Reasonable approximations to the rate of energy generation by the proton–proton chain and the carbon–nitrogen cycle are (see (4.21) and (4.22)):

$$\varepsilon = \varepsilon_0 \rho T^4, \qquad \eta = 4 \tag{5.23}$$

and

$$\varepsilon = \varepsilon_0 \rho T^{17}, \qquad \eta = 17. \tag{5.24}$$

For the four possible combinations of these laws of opacity and energy generation the constants $a_1 \ldots a_5$ have been calculated and they are shown in Table 4. In addition, the quantity $4a_3/(a_3 - 2a_1)$ which enters into the relation between luminosity and effective temperature is tabulated and it is denoted by a_6.

It should be stressed that it is quite possible that not all of these combinations of laws of opacity and energy generation will occur in stars. Each approximation is valid in some range of temperature and density as we have already seen in Chapter 4, but I do not know in advance what values of temperature and density will occur in a star of a given mass. If I calculate a series of models of stars of different masses using an opacity law (5.22) and law of energy generation (5.23), I must decide afterwards whether the physical conditions in the models are such that these laws would be valid. What I expect to find is that for stars in a certain mass range the results will be consistent; the physical conditions in the stars will be such that the laws of opacity and energy generation assumed are a good approximation. For stars outside that mass range another approximation to the laws would have been more appropriate and the calculations must be repeated.

What can be said straight away from the results of Table 4 is as follows. All of the opacity and energy generation laws predict a mass–luminosity relation with the luminosity proportional to a power of the mass between 3 and 5.5. The observational mass–luminosity relation discussed in Chapter 2 is not a simple power law but, if approximated to a power law, it has an exponent in the same range. Thus the observations suggest that for main sequence stars of about solar mass $L_s \propto M_s^5$ while for more massive stars $L_s \propto M_s^3$. In addition it may be noted that the exponent depends strongly on the law of opacity but only slightly on the law of energy generation.

In fact, the mass–luminosity relation was reasonably well understood before the law of energy generation in stars was known accurately and even before the

Table 4. *Constants occurring in homology relations for four laws of opacity and energy generation*

λ	ν	η	a_1	a_2	a_3	a_4	a_5	a_6
1	3	4	3/7	−2/7	3	4/7	2/7	28/5
1	3	17	4/5	−7/5	3	1/5	−6/5	60/7
2	6.5	4	1/13	10/13	71/13	12/13	22/13	284/69
2	6.5	17	9/13	−14/13	67/13	4/13	−10/13	268/49

nuclear origin of stellar energy had been established. The reason for this is as follows. If I assume some power law

$$\varepsilon = \varepsilon_0 \rho^\alpha T^\eta \tag{5.25}$$

for the law of energy generation, but do not know the values of α and η, homologous solutions of the equations certainly exist and a relation can be obtained between the constants a_3 and a_1 in (5.3) which does not involve α and η. Thus (5.4), (5.6), (5.10) and (5.12) combine to give:

$$a_3 = (\nu - \lambda + 1) + (3\lambda - \nu)a_1, \tag{5.26}$$

and, if this is combined with (5.3) evaluated at $m = 1$, it can be seen that

$$L_s = [\bar{L}(1)/\bar{r}(1)^{3\lambda-\nu}]M_s^{\nu-\lambda+1}r_s^{3\lambda-\nu}. \tag{5.27}$$

Equation (5.27) is a mass–luminosity–radius relation which can only be converted to a mass–luminosity relation when the law of energy generation is known so that the modified equation (5.8), with α included as well as η, can be used to enable a_1 to be eliminated from (5.26). However, for the laws of opacity (5.21), (5.22) it can be seen that $3\lambda - \nu$ is zero or small and the dependence of luminosity on radius is slight. Eddington obtained the relation (5.27) before anything was known about nuclear reactions in stars and he showed that it was in qualitative agreement with the observed mass–luminosity relation.

Effect of variation of chemical composition

I have discussed above the properties of stars of different mass and the same chemical composition. It is now possible to ask how the properties of a star of a given mass are altered if its chemical composition is changed. Provided I still assume that radiation pressure can be neglected and that the laws of opacity and energy generation are power laws, it is possible to find how the properties of a star vary with chemical composition without solving the equations completely. I will discuss this for one particular pair of laws of opacity and energy generation but the results are quite typical. The laws chosen are reasonable first approximations for stars of about a solar mass.

Suppose that the law of opacity is:

$$\kappa = \kappa_0 Z(1 + X)\rho/T^{3.5} \tag{5.28}$$

and the law of energy generation is:

$$\varepsilon = \varepsilon_0 X^2 \rho T^4, \tag{5.29}$$

where, as in Chapter 4, X and Z are the fractional abundances by mass of hydrogen and the heavy elements and where I have now made explicit the dependence of the laws of opacity and energy generation on the chemical composition of the star. Suppose also that Z is so small that a good approximation to the mean molecular weight is:

$$\mu = 4/(3 + 5X). \tag{4.47}$$

It can now be shown that, with the laws (5.28), (5.29), (3.75)–(3.78) and (5.1) have solutions of the form:

$$
\left.
\begin{aligned}
r &= r_1(X)r_2(Z)r_3(M), \\
\rho &= \rho_1(X)\rho_2(Z)\rho_3(M), \\
L &= L_1(X)L_2(Z)L_3(M), \\
T &= T_1(X)T_2(Z)T_3(M)
\end{aligned}
\right\} \tag{5.30}
$$

and

$$P = P_1(X)P_2(Z)P_3(M).$$

If this is true it follows that, if we change the chemical composition of a star of a given mass, we can predict how its properties such as luminosity, radius and effective temperature will change without solving the full equations.

The detailed argument is rather similar to that given for the case of stars of the same chemical composition and varying mass and it will not be given here. The key factor in the solution is that X and Z only occur algebraically in the equations and algebraic expressions for $r_1(X)$, $r_2(Z)$, etc. must be found so that the terms in X and Z are the same on both sides of all the equations. The explicit form of the X and Z dependence of the solutions (5.30) is given below and it can be verified that, if these expressions are substituted into (3.75)–(3.78) and (5.1), they do lead to equations for r_3, ρ_3, L_3, T_3 and P_3 in terms of M in which X and Z do not occur. The solutions are:

$$
\left.
\begin{aligned}
r &= X^{4/13}(1 + X)^{2/13}(3 + 5X)^{7/13}Z^{2/13}r_3(M), \\
\rho &= X^{-12/13}(1 + X)^{-6/13}(3 + 5X)^{-21/13}Z^{-6/13}\rho_3(M), \\
L &= X^{-2/13}(1 + X)^{-14/13}(3 + 5X)^{-101/13}Z^{-14/13}L_3(M), \\
T &= X^{-4/13}(1 + X)^{-2/13}(3 + 5X)^{-20/13}Z^{-2/13}T_3(M)
\end{aligned}
\right\} \tag{5.31}
$$

and

$$P = X^{-16/13}(1 + X)^{-8/13}(3 + 5X)^{-28/13}Z^{-8/13}P_3(M).$$

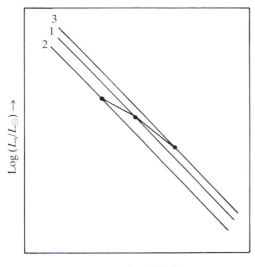

Fig. 48. Three main sequences with different chemical compositions. Sequence 2 contains more helium and sequence 3 more heavy elements than sequence 1. The filled circles show the three positions of a star of given mass.

It is also possible to combine the expressions for L and r evaluated at the surface $M = M_s$ to obtain an expression for how the effective temperature depends on the abundance of hydrogen and the heavy elements. This expression is:

$$T_e = X^{-5/26}(1 + X)^{-9/26}(3 + 5X)^{-115/52}Z^{-9/26}[L_3(M_s)/\pi a c r_3^2(M_s)]^{1/4}.$$

$$(5.32)$$

From the results (5.31), (5.32) it can be seen that if either X is increased keeping Z constant, or if Z is increased keeping X constant, or if both are increased, there is an increase in the radius of the star and a decrease in both the luminosity and the effective temperature. The changes refer to a star of a given mass. If the position of the main sequence in the HR diagram is considered it is found that when the composition is changed the displacement of the main sequence is smaller than the change in position of individual stars. This is illustrated in fig. 48. In this diagram are shown main sequences for three different chemical compositions and the positions on each of the sequences of a star of given mass. This means that homogeneous stars of different chemical compositions lie on a rather broadened main sequence.

One result of the present calculations, which proves to be of great importance when it is confirmed by calculations in which more accurate laws of opacity and energy generation are used, is that stars with a homogeneous chemical composition which are releasing energy through the conversion of hydrogen into helium are situated in the general region of the observed main sequence and they do not lie in the regions of the giants, supergiants and white dwarfs. The position of a star in the main sequence band does depend on its chemical composition, but for all possible chemical compositions there are the important regions in the HR diagram mentioned above which are not covered. This means, for example, that, if stars are well mixed as they convert hydrogen into helium so that their chemical composition remains uniform, they remain in the neighbourhood of the main sequence as they evolve. The same may not be true if, as was generally assumed in Chapter 3, the changes of chemical composition are localised where they occur so that the star becomes chemically non-uniform and this will be discussed further in the next chapter.

It should be noted that a homogeneous star which contained no hydrogen would be able to release energy by the conversion of helium into carbon. In such a star a good approximation to the law of energy release would be something like (4.28):

$$\varepsilon_{3He} \simeq \varepsilon_3 X_{He}^3 \rho^2 T^{40}. \tag{4.28}$$

Clearly homologous stellar models would also exist for such a law of energy release and for such stars there would be a helium burning main sequence which would have a different slope from the hydrogen burning main sequence. Once again detailed calculations with accurate expressions for opacity and energy release show that the helium burning main sequence would lie to the left of the hydrogen burning main sequences; stars of a given luminosity would have a higher surface temperature. I shall explain in Chapter 7 that the process of stellar mass loss can in some cases produce an approximation to a helium burning main sequence star.

All of the above results suggest that it is plausible that main sequence stars are in fact stars of homogeneous chemical composition which are gradually burning hydrogen into helium in their interiors and that the precise position of a star on the main sequence is determined primarily by its mass and to some extent by its chemical composition. The discussion given above is, however, far from conclusive as many approximations have been made in discussing these homologous models. The approximations may be listed again as follows:

   (i) Neglect of convection.
  (ii) Neglect of radiation pressure.
 (iii) Use of simple formulae for ε and κ.
  (iv) Use of the boundary condition $\rho = T = 0$ at $M = M_s$, instead of a more realistic condition.

The existence of the homologous models depended crucially on the expressions (4.39), (4.23) and (5.1) for κ, ε and P being products of powers of ρ and T, and on the boundary conditions having simple expressions at $m = 0$ and $m = 1$. Any modification of assumptions (ii), (iii) and (iv) will almost certainly mean that such homologous solutions do not exist. Thus, for example, if (5.1) is replaced by:

$$P = \frac{\mathcal{R}\rho T}{\mu} + \frac{1}{3}aT^4,$$

all the terms in this equation will only depend on the same power of M_s, if the values of λ, ν and η are such that $a_2 = 3a_4$.

If (ii), (iii) and (iv) are assumed to be valid and convection is supposed to occur only in an inner region in which all of the energy generation occurs and in which convection is so efficient that

$$\frac{P}{T}\frac{dT}{dP} = \frac{\gamma - 1}{\gamma}, \tag{3.89}$$

then homologous solutions can still be obtained. In the inner region, where (3.89) is valid, this equation takes the form

$$\frac{\bar{P}}{\bar{T}}\frac{d\bar{T}}{d\bar{P}} = \frac{\gamma - 1}{\gamma} \tag{5.33}$$

for any values of a_4 and a_5 in (5.3), and the values of a_4 and a_5 required in the outer regions of the star can also be used in the convective core. The demonstration that homologous solutions do exist is rather complicated as the inner region in which convection is occurring is surrounded by an outer region in which energy is carried by radiation and only the result can be quoted here. It is found that stars with convective cores and power laws for opacity and energy generation all have the same fraction of their mass in the convective core.

If a star has an outer convective region where, as discussed on page 77 convection is unlikely to be so efficient that (3.89) is valid everywhere, the equations governing the convective region are much more complicated and there are no longer homologous solutions even if (ii), (iii) and (iv) are assumed true. Later in this chapter I will discuss circumstances when assumption (iv) cannot be used and I will then discuss what is a more realistic boundary condition.

I have now given a long enough discussion of the properties of homologous stellar sequences, which are obtained if approximate expressions for the laws of pressure, opacity and energy generation are used. We have probably obtained a qualitative understanding of the properties of homogeneous stars but, if any quantitative comparison is to be made between theory and observation, the best possible expressions must be used for the physical quantities. As the equations of stellar structure can then only be solved on a computer, I can do no more than discuss the results which have been obtained and what, if anything, remains to be done.

General properties of homogeneous stars

The general properties of stars, which are burning hydrogen into helium and which are chemically homogeneous either because the burning has only just started, or because for some reason they are well mixed, are as follows:

(i) High mass stars have a region in the centre in which a significant proportion of the energy is carried by convection. The central temperature is an increasing function of stellar mass, as was suggested by the homology results of Table 4, and this means that in the high mass stars energy generation will be by the CN cycle rather than the PP chain. In addition, as predicted by most of the homologous results, the central density decreases with increasing stellar mass.

From fig. 43 in Chapter 4 it can be seen that at sufficiently high temperatures an increase in temperature, or a decrease in density, tends to make the opacity scarcely dependent on temperature and density; as has been stated on page 101, this occurs when electron scattering is the main source of opacity. This means that the second row of Table 4 (constant opacity and CN energy generation) is likely to give the best approximation to the properties of these stars. Convection occurs in the central regions of these stars because the CN cycle provides a highly concentrated source of energy and radiation alone is not adequate to carry the energy away from the central regions. The existence of a convective core in these stars implies that the material in the deep interior is well mixed and this in turn, as we shall see in the next chapter, has important consequences for their evolution.

(ii)   In lower mass stars, the main energy generation is by the PP chain and the opacity is more nearly due to Kramers' law:

$$\kappa = \kappa_2 \rho / T^{3.5}. \tag{5.22}$$

The source of nuclear energy is no longer sufficiently concentrated to give rise to a convective core. These stars do, however, have outer regions in which convection occurs. This can be explained as follows. The surface temperatures of high mass main sequence stars are higher than those of low mass stars as has been predicted by the homologous solutions in Table 4 and they are such that the abundant elements, hydrogen and helium, are ionised at the stellar surface. In the low mass stars the surface temperatures are lower and these elements are neutral. In that case, there is a region just below the stellar surface in which these elements are being ionised and in these ionisation zones the ratio of specific heats of the stellar material is very much smaller than usual and can be comparable with unity. As a result the criterion for convection

$$\frac{P}{T}\frac{\mathrm{d}T}{\mathrm{d}P} > \frac{\gamma - 1}{\gamma} \tag{3.68}$$

is satisfied. In these outer convection zones the structure of the star does depend on the efficiency with which convection can carry energy and the lack of a really good theory of convection means that there is an important uncertainty in the theory. *This is probably the biggest single uncertainty in the theory of main sequence stars being more important than inaccuracies in the laws of opacity and energy generation.* We can turn the discussion round and ask how much energy must be carried by convection if theory and observation are to agree and the most that can be said is the the quantity does not seem implausible.

(iii)   There is a mass below which no consistent equilibrium solution of the equations can be obtained. As has been mentioned earlier, the central temperature is an increasing function of stellar mass while the opposite is true of the central density. This means that as the stellar mass is reduced, there is an increasing tendency for the material in the centre of the star to become a degenerate gas; we have seen in Chapter 4 that the ideal classical gas law breaks down at low temperature and high density. When the evolution of very low mass stars towards the main sequence is studied it is found that, although there is initially an increase in the central temperature, it reaches a maximum value and decreases again when the centre becomes a degenerate gas before any significant nuclear reactions converting hydrogen and helium take place. Such stars do have a luminous phase in which they are contracting and releasing gravitational energy but this is very much shorter than the main sequence lifetimes which they would have if they were just a little more massive. Such stars cool down and die without ever *tapping* a nuclear energy source and without having a main sequence phase. The critical mass below which this happens depends slightly on the chemical composition of the star and it is about $0.08M_\odot$. Such stars have been given the name *brown dwarfs*. They are very difficult to detect even if they are very close to the Earth and there continues to be dispute about whether they exist in large numbers. Theories of star formation do not rule out the formation of stars of such low mass and they may be an important form of invisible matter in the Universe.

The critical mass depends fairly slightly on chemical composition provided that the main sequence stars are obtaining their energy by the burning of hydrogen into helium. If calculations are made for stars which contain no hydrogen and which are mainly composed of helium so that the main source of energy is helium burning, the lowest mass for which a consistent main sequence model can be obtained is about $0.35M_\odot$, for a pure carbon star it is even higher at about $0.8M_\odot$. As present evidence suggests that all main sequence stars do contain a considerable proportion of hydrogen, these latter results may be of only theoretical interest unless it is possible for stars to lose all of their hydrogen at a later stage of evolution and become pure helium stars.

(iv)   The qualitative results concerning the existence of a main sequence and a mass–luminosity relation and the dependence of the position of the main sequence on chemical composition found for homologous stars are confirmed by more careful calculations. The main sequence is no longer found to be precisely a straight line and the mass–luminosity relation is no longer a pure power law. It is also true that the observed relations between mass and luminosity and luminosity and effective temperature are not simple power laws.

(v)   Because the luminosity of a star depends on a fairly high power of its mass ($L_s \propto M_s^3$ or $\propto M_s^5$ depending on mass range) while the nuclear energy which can be released in converting a given fraction of its mass from hydrogen into helium is directly proportional to the mass, the time for which stars can exist in a hydrogen burning phase is smaller for stars of greater mass. Thus the total energy release in converting hydrogen into helium if a star is initially composed of pure hydrogen and all of the hydrogen is burnt is:

$$E_{\text{H}\rightarrow\text{He}} = 0.007M_s c^2. \tag{5.34}$$

This will be an over-estimate of the actual amount of energy released in the hydrogen burning phase both because a star is unlikely to be made of pure hydrogen and because, as we shall see later, 100% efficiency in the conversion of hydrogen to helium is unlikely. If the main sequence luminosity when the star is made of pure hydrogen is L_s, an estimate of the time for which hydrogen burning can occur (the *main sequence lifetime* of the star) is:

$$t_{\text{H}\rightarrow\text{He}} = 0.007M_{\text{s}}c^2/L_{\text{s}}. \tag{5.35}$$

This is likely to be an over-estimate of the time, not only for the two reasons listed above, but also because L_{s} will increase as the conversion of hydrogen into helium proceeds. I have already shown that (5.31) predicts this property for stars which remain homogeneous as they evolve. I shall show in the next chapter that the same is also true for stars which develop non-uniformities of chemical composition as they evolve.

Estimates of the main sequence lifetimes of stars of different masses, based on calculations of the evolution of hydrogen burning stars are shown in Table 5. These times are shorter than those obtained by a simple use of (5.35) because allowance has been made for the factors causing (5.35) to be an over-estimate. One very important result can be deduced from this table. It is known, from the properties of the radioactive elements in the Earth's crust, that the Earth has been solid for about 4.5×10^9 years. From Table 5 it is clear that *main sequence stars with a mass greater than about* $1.25M_\odot$ *cannot have been on the main sequence when the Earth solidified.*

As the first important release of nuclear energy occurs when stars are on the main sequence, the pre-main-sequence lifetime (which will be discussed briefly later in this chapter) is thought to be governed by the release of gravitational energy giving a lifetime of order 3×10^7 years for the Sun (equation (3.40)) and a lifetime of less than 4.5×10^9 years for all stars in the mass range which is observed. Thus not only can we say that massive stars cannot have been on the main sequence all of the time since the Earth solidified, but that any massive main sequence star which we observed today must have been formed long since the Earth solidified. It can be seen from the Table that stars more massive than $15M_\odot$ have probably been formed in the last ten million years. This is perhaps the first definite evidence that not all astronomical objects came into existence at the same time and it makes it very likely that genuine new stars (as opposed to novae which are old stars becoming brighter) are being formed today.

It should be stressed that these new stars are not being created out of nothing. Most astronomers believe that our Galaxy started forming out of intergalactic gas between 10^{10} and 2×10^{10} years ago. During the time that has passed since then, stars have formed out of condensations in this gas. What I am saying in this section is that all of the stars were not formed at once, but that there is evidence that there has been a continuing process of star formation throughout the lifetime of the Galaxy. As there is still interstellar gas and dust in the Galaxy, raw material is available for the formation of new stars today. From Table 5 it can be deduced that stars of one solar mass and less which formed quite early in the lifetime of the Galaxy are still luminous today whereas massive stars formed early in the galactic lifetime must have completed their life history long ago. Whenever a star loses mass to the interstellar medium, as occurs in the explosion of novae and supernovae, this material can then be incorporated in a future generation of stars. Some star clusters, whose HR diagrams have been discussed in Chapter 2, contain very luminous, and hence presumably massive, stars on their main sequences. If the idea expressed in that chapter that all stars in a cluster have essentially the

Table 5. *Main sequence lifetime (in years) for stars of different masses*

$M/M_\odot$	15.0	9.0	5.0	3.0	2.25	1.5	1.25	1.0
Lifetime	1.0×10^7	2.2×10^7	6.8×10^7	2.3×10^8	5.0×10^8	1.7×10^9	3.0×10^9	8.2×10^9

same age is true, this means that *entire clusters of stars have been formed quite recently in the lifetime of the Galaxy*. This will be discussed further in the following chapter.

(vi) As mentioned earlier, whatever chemical composition and set of energy-releasing nuclear reactions is assumed for the stars, there are regions in the HR diagram which are not covered by the set of possible main sequences. *Giants, supergiants and white dwarfs cannot be chemically homogeneous stars in which important nuclear energy release is taking place.* It is believed that giants and supergiants have important non-uniformities of chemical composition and that there is no significant nuclear energy release in white dwarfs. The structure of giants and supergiants will be discussed in Chapter 6 and that of white dwarfs in Chapter 10

Detailed calculations of main sequence structure and comparison with observation

In any really detailed comparison of the results of this section with observation there are a variety of difficulties. There are inaccuracies in the theoretical calculations because of uncertainties in such quantities as the opacity, energy generation and transport of energy by convection. There are the problems discussed in Chapter 2 of converting the observed magnitudes and colour indices into the bolometric magnitude and effective temperature calculated by the theorist. Quantities such as mass and radius can only be observed directly for a limited number of stars. Finally, only the chemical composition of the outer layers of a star can be deduced from observations and there is no direct evidence about whether or not a star has a homogeneous chemical composition. For these reasons the results obtained by different authors do not agree in fine detail even though it is believed that the main sequence phase of stellar evolution is generally well understood. Some results of calculations are described below.

The main sequences for four different chemical compositions which were at the time of calculation thought to be appropriate to stars in the solar neighbourhood in the Galaxy are shown below in fig. 49. The compositions are:

$$
\begin{aligned}
&(a) &&X = 0.60, &&Y = 0.38, &&Z = 0.02,\\
&(b) &&X = 0.70, &&Y = 0.28, &&Z = 0.02,\\
&(c) &&X = 0.60, &&Y = 0.36, &&Z = 0.04,\\
&(d) &&X = 0.70, &&Y = 0.26, &&Z = 0.04.
\end{aligned}
\qquad (5.36)
$$

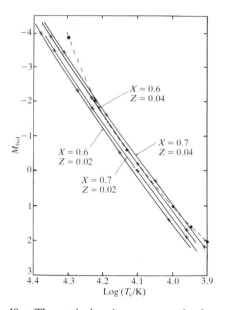

Fig. 49.   Theoretical main sequences for four chemical compositions. Also shown as a dashed curve is a section of the observed main séquence.

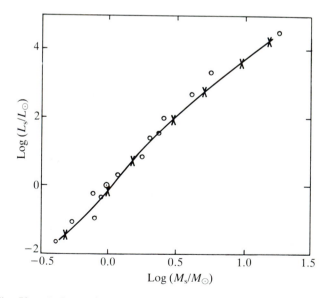

Fig. 50.   A theoretical mass-luminosity relation based on calculations by Iben. Also shown are some main sequence stars of well-determined mass and luminosity including the Sun ⊙.

Table 6. *Fractional mass in convective core (M_{cc}) or convective envelope (M_{ce}) for main sequence stars of different masses*

$M/M_\odot$	15.0	9.0	5.0	3.0	1.5	1.0	0.5
M_{cc}	0.38	0.26	0.21	0.17	0.06	0.00	0.01
M_{ce}	0.00	0.00	0.00	0.00	0.00	0.01	0.42

It can be seen that the calculated main sequences are approximately straight lines with a slope in general agreement with the observed main sequence band. Also shown in the diagram is a mean line through the main sequence deduced from observations of nearby stars with appropriate conversion from observed quantities to $\log L_s$ and $\log T_e$. Of course the observed main sequence band is quite broad and all of the above lines are within the region of the observed main sequence. It can be noted from fig. 49 that the position of the main sequence is particularly sensitive to a change in the heavy element content, Z, and this is in general agreement with the homology result of (5.32). Since these calculations were performed, it has become clear that $X = 0.7$ is much more appropriate than $X = 0.6$ but it can be seen that the main sequence position is not greatly affected by such a change in X.

For a slightly different chemical composition than any of the above:

$$X = 0.71, \qquad Y = 0.27, \qquad Z = 0.02 \tag{5.37}$$

I. Iben has calculated main sequence models for a variety of masses between $0.5M_\odot$ and $15M_\odot$ and he has also studied both pre- and post-main-sequence evolution of these stars. From his main sequence results it is possible to draw up a theoretical main sequence mass–luminosity relation and that is shown in fig. 50. Also shown in the diagram are points corresponding to nearby stars of known mass and luminosity and it can be seen that once again there is a good qualitative agreement between theory and observation. Table 6 shows what fraction of the mass of each star is in either a convective core or a convective envelope.

It was mentioned in Chapter 2 that there is a group of stars known as sub-dwarfs which lie below the main sequence in the HR diagram. It was also mentioned that the sub-dwarfs appeared to have a low content of heavy elements. It can be seen from the results given above that stars with low Z do lie below stars with high Z and the low content of heavy elements may go at least some way towards explaining the position of the sub-dwarfs in the HR diagram.

Pre-main-sequence evolution

I will now give a brief description of what is at present known about pre-main-sequence evolution. As mentioned in the introduction to this chapter, the

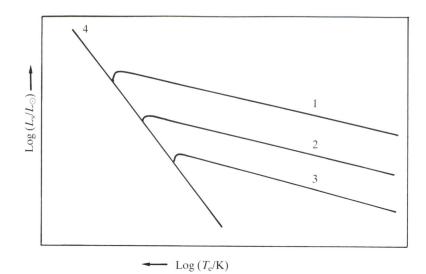

Fig. 51.   The approach to the main sequence of fully radiative stars. The curves 1, 2, 3 refer to stars of three different masses and the line 4 is the main sequence.

subject of star formation and pre-main-sequence evolution is at present not very well understood and much of what will be said in the remainder of this chapter would be disputed by some workers in the field. In the first instance if, as is believed, stars have been formed out of condensations in the interstellar gas, their initial state must be one of very large radius and very low luminosity and surface temperature; that is, a newly formed star is low down to the right in the HR diagram. In my initial discussion I shall assume that to a good approximation such protostars are spherical and non-rotating but I shall comment on that approximation later. When the pre-main-sequence evolution of stars was first studied it was assumed that the evolution to the main sequence was a process in which the star's radius decreased and the luminosity and surface temperature increased steadily. Calculations in which it was assumed that all of the energy transport in protostars was by radiation gave results with essentially that character and these are shown in fig 51. The surface temperature increases steadily and the radius decreases steadily but there is a maximum in the luminosity just before the main sequence is reached. This occurs because, when nuclear reactions commence in the central regions of the stars, they lead to a rise in temperature and pressure in the central regions. This is followed by a small expansion in the central regions and a decrease in temperature and luminosity.

When the properties of these contracting protostars were studied in detail, it was found that they had some defects. In the first instance it was found that they contained important regions in which the criterion

$$\frac{P \, dT}{T \, dP} > \frac{\gamma - 1}{\gamma},$$

(3.68)

for convective instability *was* satisfied. The assumption that all of the energy was carried by radiation was incorrect. In addition it was found that use of the boundary condition

$$\rho = 0, \qquad T = 0 \quad \text{at} \quad M = M_s \tag{3.80}$$

was inadequate.

Improved surface boundary condition

The visible surface of a star is the level from which radiation can just about escape without any further absorption. The temperature of the visible surface has been assumed to be approximately T_e; that is the meaning of the equation:

$$L_s = \pi a c r_s^2 T_e^4. \tag{2.7}$$

A solution of the stellar structure equations obtained using the approximate surface boundary condition (3.80) will be a good solution provided radiation can just about escape from the level where $T = T_e$. If we suppose that in such a solution $T = T_e$ at $r = r_e$ while $T = 0$ at $r = r_s$, radiation will just about escape from the level where $T = T_e$ if

$$\int_{r_e}^{r_s} \kappa \rho \mathrm{d}r \simeq 2/3. \tag{5.38}$$

This result is obtained as follows. Equation (4.29) can be integrated to show that radiation travelling radially outwards in a star is attenuated exponentially so that

$$(I_\nu)_{r=r_s}/(I_\nu)_{r=r_e} = \exp\left[-\int_{r_e}^{r_s} \kappa \rho \, \mathrm{d}r\right]. \tag{5.39}$$

If (5.38) is satisfied about half the radiation is absorbed and half escapes. For many stars (5.38) is satisfied by solutions of the equations in which the simple boundary condition (3.80) has been used. However, for protostars, and in fact any stars which have deep outer convection zones, it is necessary to use a more accurate boundary condition based on (5.38). It is possible to simplify (5.38) by using approximate forms of the equations of stellar structure valid near the surface of the star. I will not give the argument but the final answer is that (5.38) can be replaced by

$$P\kappa = 2GM/3r^2 \quad \text{at} \quad M = M_s. \tag{5.40}$$

This is then one boundary condition at the surface of the star and the second one can be taken to be

$$L = \pi a c r^2 T^4 \quad \text{at} \quad M = M_s, \tag{5.41}$$

in effect (2.7).

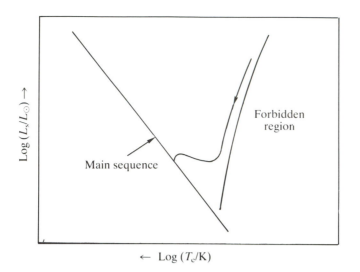

Fig. 52.   The final approach to the main sequence according to Hayashi.

Hayashi's theory of pre-main-sequence evolution

When Hayashi introduced the better surface boundary condition and allowed for the existence of energy transport by convection, he found that convection could be important throughout the whole of a protostar. More unexpectedly he found that there was a region in the HR diagram in which he could find no equilibrium solutions to the equations of stellar structure. This is known as the *Hayashi forbidden zone* and it is shown in fig. 52. Also shown is his result for the immediate pre-main-sequence evolution of a star in which it can be seen that he predicts that stars can have a pre-main-sequence luminosity much higher than their main sequence luminosity. Stars just at the boundary of the Hayashi forbidden zone have convection occurring throughout their interiors and stars to the left of the forbidden region have some regions in which convection is not occurring.

If the immediate pre-main-sequence luminosity can be much higher than the main sequence luminosity, it is important to try to understand why this is so and how the immediate pre-main-sequence state can be reached from the initial state of low luminosity and surface temperature. Hayashi's work gives a mathematical answer but a physical understanding is also desirable. In the early stages of the evolution of a protostar its luminosity is determined by how rapidly it can radiate energy and there is no reason why this rate of loss of energy should bear any relation to its main sequence luminosity. In the earliest stages the stellar material is transparent rather than opaque to radiation and the luminosity increases rapidly as the star contracts. Eventually the star becomes opaque and traps radiation within it and at this stage there is a decrease in luminosity. Throughout these initial stages the stellar material has radiated energy so efficiently that its temperature has probably been between 10K and 20K.

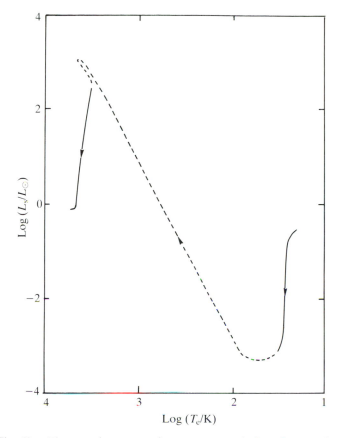

Fig. 53. The complete pre-main-sequence evolution of a star of solar mass.

Once the star is opaque its internal temperature rises and there comes a time when further increase of temperature causes first molecular hydrogen to be dissociated and then atomic hydrogen to be ionised. Both these changes of state require a considerable amount of energy which can only come from the gravitational energy of the star and they trigger off another stage of rapid contraction in which there is another substantial increase in luminosity. Finally when essentially all of the material is ionised the rapid collapse stops and the star enters the final approach to the main sequence shown in fig. 52.

One calculation of the approach to the main sequence for a star of solar mass has been made by Hayashi and his colleagues and it is illustrated in fig. 53.† Although this is a fairly detailed discussion of pre-main-sequence evolution it cannot be regarded as a conclusive study. Other workers have made calculations which give results which are very different from those of Hayashi. Interstellar gas

---

† It appears that the results of fig. 53 contradict the existence of the Hayashi forbidden zone. However, Hayashi's result was obtained on the assumption that very rapid time variations were not occurring and they do occur in all the early phases shown in fig. 53.

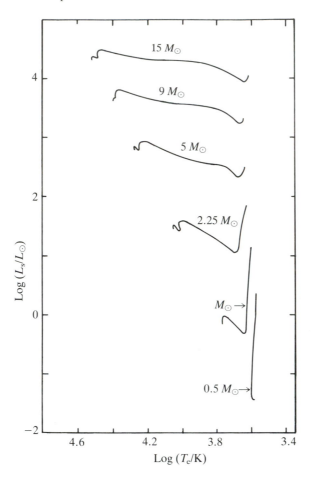

Fig. 54.   The final approach of stars of various masses to the main sequence.

clouds are observed to have rotational energy and to be pervaded by interstellar magnetic fields and both of these properties may seriously affect star formation and pre-main-sequence evolution. In addition some stars which are believed to be approaching the main sequence are found to be variable stars of the T Tauri type. These are irregular variables which are losing mass to the interstellar medium and such mass loss may play an important role in pre-main-sequence evolution. I will comment on some of these points after I have completed the simplest discussion.

Some calculations of the approach to the main sequence due to Iben are shown in fig. 54. It can be seen that the final stage of the approach to the main sequence is similar to the results shown for completely radiative stars in fig. 51, provided only that the main sequence position of the star is well to the left of the Hayashi forbidden region. This is true for massive stars and for them the pre-main-sequence luminosity peak is certainly less significant than for low mass stars.

The whole of pre-main-sequence evolution is rapid compared to the main sequence lifetime. In (3.41) I estimated how long the Sun could have radiated at

Table 7. *Time taken to reach main sequence for stars of different masses (mass in $M_\odot$ and time in years*

Mass	15.0	9.0	5.0	3.0	2.25
Time	6.2×10^4	1.5×10^5	5.8×10^5	2.5×10^6	5.9×10^6
Mass	1.5	1.25	1.0	0.5	
Time	1.8×10^7	2.9×10^7	5.0×10^7	1.5×10^8	

its present rate if its only energy supply had been from gravitational contraction and obtained the result:

$$t \simeq GM_\odot^2/L_\odot r_\odot \simeq 10^{15} \text{ s} \simeq 3 \times 10^7 \text{ years.} \tag{3.43}$$

At that stage I made this estimate to demonstrate that the Sun must have had a different energy source during the past 10^9 years. During the pre-main-sequence phase the main source of energy is gravitational contraction so a first estimate of how long it took the Sun to reach the main sequence is given by (3.43). Now it is believed that the Sun may have been 500 times as luminous as it is now at some stage in the past, it seems possible that the pre-main-sequence age could be significantly less than the value given in (3.43). In fact, for stars of a solar mass and greater, this reduction does not occur. It has been realized that the initial reactions of the CN cycle (4.16), which convert ^{12}C into ^{14}N, occur before the main sequence is reached and increase the time of approach to the main sequence. The effect of these reactions on the overall chemical composition of a star is slight and this does not affect our assertion that stars reach the main sequence with essentially their original chemical composition.

The calculated times to reach the main sequence for Iben's models are shown in Table 7. For massive stars the time to reach the main sequence is longer than that given by the analogue of (3.43) because the additional time allowed by the nuclear reactions between the light elements is greater than the reduction due to the high luminosity phase. For the star of one solar mass, Hayashi's calculations show that the entire evolution before the maximum luminosity is reached lasts only about 20 years. Because the pre-main-sequence evolution is very rapid compared to the main sequence lifetime, we are likely to observe only a relatively small number of stars in the stage of pre-main-sequence contraction if stars have been forming steadily during the lifetime of the Galaxy. Thus, in this case, *the number of stars observed in a given phase of evolution should be roughly proportional to the time an individual star spends in that phase.* The simplest example of this is that most stars are main sequence stars because stars spend a large proportion of their life on the main sequence. This will be discussed in Chapter 6. It seems very unlikely indeed that we shall be able to identify any stars as being in the stage of evolution before the final luminosity maximum, although one such identification has been tentatively suggested. This is one of the basic difficulties with trying to observe stellar evolution. *Normally evolution occurs so very slowly that observation of it is quite impossible. When evolution is rapid the phase of evolution is soon over and then the statistics are against our finding any stars in that phase.* One exception, where rapid evolution is observed, is the explosion of a supernova. These stars become so highly luminous that they draw attention to themselves even at very large distances.

One consequence of our present views on the pre-main-sequence evolution of the Sun should be mentioned. It is thought that the Earth and the other planets were formed out of solar material or at least out of the same material as that from which the Sun was formed. Until Hayashi's work it was always assumed that during the period of planetary formation the Sun was radiating less energy than it is today. Now it seems likely that the Sun was more luminous for part of the time and this might have important effects on theories of the origin of the solar system.

Complications concerning pre-main-sequence evolution

I will now comment on a variety of possible complications to the simplest ideas about pre-main-sequence evolution. Ideally a theoretical study would start with an interstellar gas cloud and follow its evolution as protostars formed in it and then contracted to form main sequence stars. This is a very difficult problem both physically and mathematically with the density of a typical star being more than a factor 10^{20} higher than a typical gas cloud. In practice pre-main-sequence evolution has been studied by assuming that a protostellar fragment exists which is of some density intermediate between that of a gas cloud and of a star. In the earliest calculations, including those of Hayashi, the protostars were assumed to be spherical, non-rotating and of uniform density. All of these assumptions are likely to be incorrect.

It is not difficult to consider the case in which the central region of the protostar is rather more dense than the outside. This is not *a priori* improbable as one might expect that protostars are formed around a local peak in density in an interstellar cloud. Whereas a contracting uniform protostar remains of uniform density until it becomes opaque, the denser central regions of a non-uniform cloud collapse more rapidly than the outside and this allows the possibility that part of the protostar could reach stellar densities and temperatures and, indeed, start burning nuclear fuel while the outer layers were still falling in. In such a case what a protostar might look like would be very different from what is predicted by the simple discussion of pre-main-sequence evolution just given. Initially the radiation from a small star embedded in a cloud of infalling gas could be absorbed and re-radiated in the infrared region of the spectrum. At some stage the cloud would become partially transparent and the central star would become visible inside the infrared source. Finally, when all the infalling matter was accreted, an ordinary main sequence star would result. Observations have suggested that this is a relevant evolutionary sequence for some stars.

Something more dramatic may happen in the case of very massive stars. If the initial central star is sufficiently massive it radiates strongly in the ultraviolet region of the spectrum. This ultraviolet radiation can ionise the infalling hydrogen

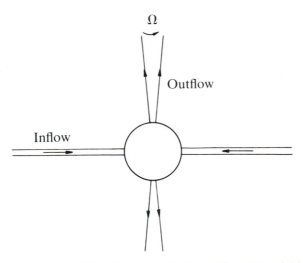

Fig. 55. A protostar with a disk and a bipolar outflow. Material in a disk rotating around the star moves radially and is gradually accreted. At the same time mass loss from the star is channelled in a bipolar outflow by the star's magnetic field.

and also give it enough energy that inflow is turned into outflow. The mass of the star which is formed is in that case less than that of the initial protostar. It has been suggested that there is a maximum mass for main sequence stars because if more massive stars tried to form the excess mass would be expelled. Whether or not this is true is not clear because there may be other ways in which the material of massive stars can accumulate.

Observations indicate that the assumption of a spherical non-rotating star is also seriously wrong in many cases. The subject is rather too complicated for this book but I will make a few comments. Observations have shown that some protostars are surrounded by flat disks and that they are also indulging in bipolar outflows of the form indicated in fig. 55. If a protostar is rotating, its contraction will be accompanied by flattening because of centrifugal force acting in its equatorial plane so the presence of disks around protostars is not unexpected. The planetary system is believed to have formed in such a disk around the protosun. In Chapter 8 I shall describe how such disks can exist in close binary systems around one of the stars and how the material in the disk can gradually accrete on the star which it surrounds. In the case of a protostar, the mass of the star may gradually increase by accretion of matter from the disk rather than the whole star being formed in a single discrete event. At the same time the star may be losing mass from its surface in the same way as the Sun does all the time through the solar wind which I shall discuss at the start of Chapter 7. If the protostar is magnetised like the Sun, the mass loss may be channelled by the magnetic field and produce the observed bipolar outflows.

The subject of star formation and pre-main-sequence evolution is complex. It is advancing rapidly largely because of developments of infrared and millimetre wave astronomy because much of the radiation from star forming regions and protostars is at these wavelengths. It is however likely to be a long time before the subject is fully understood. One further fact which should be mentioned is that many if not most stars are partners in binary systems so that an understanding of the formation of stars must include the formation of binary stars.

Summary of Chapter 5

In this chapter I have investigated the hypothesis that main sequence stars are stars of uniform chemical composition which are converting hydrogen into helium in their interiors. Using simple approximations to the laws of opacity and energy generation, which were discussed in Chapter 4, I have shown that such stars obey a mass–luminosity relation similar to that of main sequence stars and that they lie in a region in the HR diagram in qualitative agreement with the observed main sequence. Results of more detailed calculations, with more accurate mathematical expressions for the laws of opacity and energy generation, confirm these qualitative results. Red giants and white dwarfs are not stars of uniform chemical composition which are radiating energy which has been released in nuclear reactions.

Main sequence stellar structure is essentially independent of the star's previous life history and this is fortunate as the theory of star formation and pre-main-sequence stellar evolution still contains many uncertainties. Stars must initially be large, cool and of low luminosity and in approaching the main sequence they must become smaller, hotter and brighter. It is now believed that the luminosity of a star may have a maximum value in the pre-main-sequence phase which is much greater than its main sequence value. However the actual way in which a star forms and reaches the main sequence may be much more complicated than the simple symmetrical models which have first been calculated.

6

Early post-main-sequence evolution and the ages of star clusters

Historical introduction

After the main sequence the most prominent group of stars in the HR diagram (fig. 56) is the red giants and supergiants. These stars have larger luminosities and radii than main sequence stars of the same colour. From the discussion in Chapter 5, it appears that red giants are not stars of homogeneous chemical composition and I must now discover how red giants differ from main sequence stars in their internal structure as well as in their surface properties. I have already indicated at the end of Chapter 2 (fig. 30) that stars become red giants when nuclear reactions in their interiors lead to a non-uniformity of chemical composition. Before I discuss this further, I will give a brief historical introduction to the problem of the red giants. Although in this book I mainly discuss the present state of knowledge, it is perhaps instructive in one case to trace the steps by which the present knowledge has been obtained.

When the first theoretical calculations of stellar structure were made, it was very difficult to explain the occurrence of red giants, since at the time it was believed that stars remained chemically homogeneous as they evolved. As will now be described, it was believed that the rotation of stars caused them to be well mixed. Most stars are observed to rotate, even if the rotation of many of them is not sufficiently rapid to distort their structure substantially. Rotation is detected by the Doppler effect. As has previously been mentioned, radiation from a source moving away from an observer is shifted to the red while that from a source moving towards the observer has a blue shift. If a star is rotating, part of it is moving towards us and part away from us and this causes any spectral line from the star to be broadened (fig. 57). The observation of broadened spectral lines leads to the deduction that stars are rotating.

A rotating star is not spherical and the surfaces of constant temperature, density and pressure in such a star are spheroids to a first approximation (fig. 58). In this figure it can be seen that the temperature gradient near the poles of such a star is

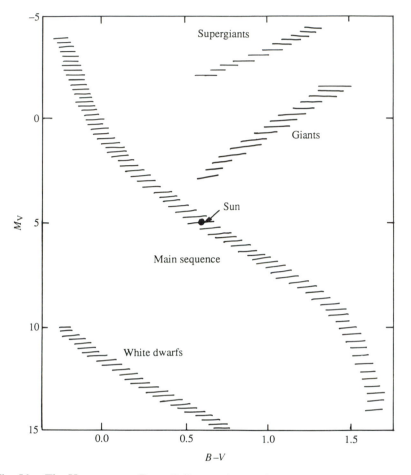

Fig. 56.   The Hertzsprung–Russell diagram for nearby stars.

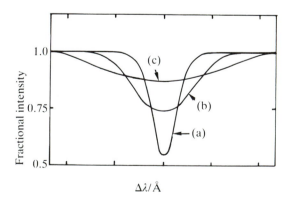

Fig. 57.   The effect of rotation on the shape of a spectral line. (a) is the shape of the spectral line, in a non-rotating star, (b), in a star of moderate rotation and (c), in a rapidly rotating star.

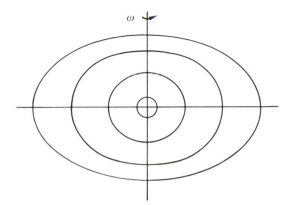

Fig. 58.   Rotating star. The surfaces shown have constant pressure and temperature.

greater than the temperature gradient near the equator as the surfaces of constant temperature are closer at the pole. This in turn means that more energy is carried by radiation near the pole and left to itself this would disturb the surfaces of constant temperature. Eddington showed that the surfaces of constant temperature are preserved by slow circulating motions of the type shown in fig. 59 which carry both material and energy from one part of a star to another. These are known as *meridional currents* and Eddington believed that they kept a star chemically homogeneous as it evolved. The typical speed of the currents is now believed to be:

$$v \approx (\omega^2 r_s/g)(L_s/M_s g), \tag{6.1}$$

where ω is the angular velocity of the star and g the acceleration due to gravity, but Eddington's original estimate was several orders of magnitude greater than this. As a result, Eddington believed that even slowly rotating stars would be well mixed. Although the meridional circulation produces some of the same effects as convection it is qualitatively rather different; it is produced by rotation rather than by a large temperature gradient and the motions are regular on a very large scale, while convective motions are very irregular. In addition, in any region in which convection is carrying energy, the convective motions are very much more rapid than the meridional circulation.

   If hydrogen burning stars remained chemically homogeneous as they evolved they would remain in the neighbourhood of the main sequence. As I have shown in Chapter 5 (fig. 48) the conversion of hydrogen into helium is accompanied by a motion to the left and upwards in the HR diagram and not into the region occupied by the red giants. As red giants did not appear to occur naturally in the process of stellar evolution, theoretical astrophysicists experimented with models which might have *giant* properties. They found that models with a single discontinuity of chemical composition (fig. 60), with the inside region of higher molecular weight, could have large radii provided that the ratio of masses in the two zones was chosen carefully. It was then necessary to think of ways in which this discontinuity of chemical composition could be produced.

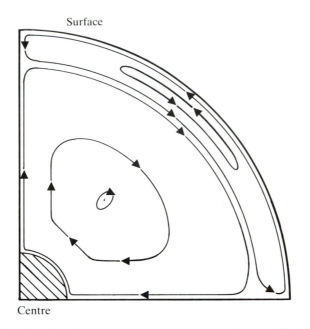

Surface

Centre

Fig. 59.   Meridional circulation currents in a rotating star. The hatched area is a convective core and the arrows show the direction of the currents.

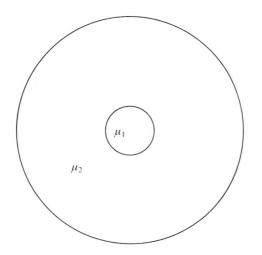

Fig. 60.   Star with discontinuous chemical composition. A region with molecular weight μ_1 is surrounded by a region with molecular weight μ_2.

F. Hoyle and R. A. Lyttleton proposed that a star passing through an interstellar gas cloud could increase its mass by *accretion* of some of the material of the cloud. If the star had already converted some of its hydrogen into helium and if the interstellar gas cloud consisted mainly of hydrogen, a discontinuity of chemical composition would be produced and the star might become a red giant. It would then remain a red giant until mixing currents once again made the star homogeneous. According to this picture a star would spend a limited time as a red giant and it might become a red giant several times in its life history.

At the time that this suggestion was made, the clouds of neutral (un-ionised) hydrogen in the Galaxy had not yet been discovered. They were subsequently detected by radio astronomers because even very cold hydrogen emits radio waves at a wavelength of 21cm (0.21m). When the distribution of neutral hydrogen clouds in the Galaxy was mapped by the radio astronomers, it was found that these clouds were neither dense enough nor slowly moving enough† to permit substantial accretion of interstellar matter by stars. In addition, it was difficult to see how the rather well-defined giant branches of galactic and globular clusters could be produced if matter was being accreted by stars of all masses in the clusters. It was fortunate that, just at the time that accretion theory became untenable, an error was discovered in the original estimates of the speed of meridional circulation currents made by Eddington. With the revised speeds, it appears that most stars will not be mixed by meridional circulation. For example, in the interior of the Sun the circulation currents are believed to travel at 10^{-11} ms^{-1} and at this speed they take more than 10^{12} years for one circulation, while substantial changes of chemical composition, due to nuclear reactions, can occur in the solar interior in less than 10^{10} years. It thus appears that natural inhomogeneities of chemical composition can arise as a star evolves and can lead to its becoming a giant. This concludes this historical introduction and I now describe our present state of knowledge.

General character of post-main–sequence evolution

I shall show that the details of post-main-sequence evolution depend on stellar mass and, in particular, on whether or not a star has a convective core when it is on the main sequence. If a main sequence star has a convective core, material from throughout the convective core can be carried into the central regions and is available for nuclear reactions, regardless of the temperature difference which exists between the centre and the surface of the core. If the star has a core in which all of the energy transport is by radiation, there will be no mixing processes in the interior of the star. Thus the exhaustion of the central supply of hydrogen will be determined entirely by the speed of nuclear reactions at the centre and not by the rate at which unburnt hydrogen can be carried into the centre.

Broadly speaking the first crucial stage in the evolution of stars away from the main sequence occurs when the central hydrogen content falls to zero. Before this

† The cloud densities can be deduced from a measurement of the total 21 cm emission from a given volume while the cloud velocities are found from the Doppler effect on the 21 cm line.

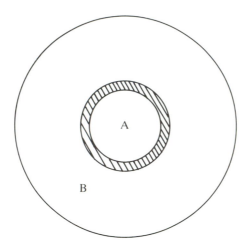

Fig. 61.   Star with a hydrogen burning shell. In region A all of the hydrogen has been converted into helium, in region B no nuclear reactions have occurred and in the hatched area hydrogen is burning into helium.

stage, calculations show that the star's luminosity gradually increases but that it remains in the general neighbourhood of the main sequence. Once there is no hydrogen left in the centre, the central regions stop releasing nuclear energy. Instead they resume their release of gravitational energy and begin to contract, slowly in the first instance. As this happens, the region in which hydrogen is being burnt gradually moves outward through the star and what is known as a hydrogen burning shell is produced (fig. 61). As the only energy release in the central regions of the star is now gravitational potential energy, the luminosity in that region is very low. Because the luminosity is low, only a very small temperature gradient is needed to carry the energy outwards (see (3.54) for example) and the star has an almost *isothermal core* composed of helium and a small admixture of heavy elements, corresponding to the initial heavy element content of the star.

As the evolution of the star is followed, it is found that a dramatic change occurs when the isothermal core contains between 10% and 15% of the mass of the whole star. It is found that the pressure gradient in a larger isothermal core, which is contracting slowly, is unable to support the outer regions of the star and the central regions now collapse rapidly instead of contracting slowly. This critical mass for a slowly contracting isothermal core is known as the *Schönberg–Chandrasekhar limit* and we shall see later in this chapter that it is very important in the understanding of the galactic cluster HR diagrams.

When calculations are continued past the stage at which the Schönberg–Chandrasekhar limiting mass is reached, it is found that the inner layers of the star contract rapidly and heat up, but that simultaneously the radius of the star as a whole expands. The heating of the inside is produced by the rapid release of gravitational energy. Although the surface expansion which accompanies the core contraction is predicted by the solutions of the equations of stellar structure for a

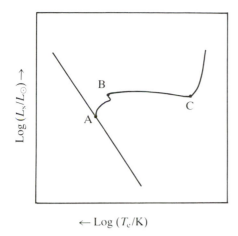

Fig. 62. Evolution to the giant region. An isothermal core is formed near B and a deep outer convection zone appears near C.

star with variable chemical composition and although such an expansion is necessary to explain the existence of red giants, it is not easy to give a simple explanation of why it occurs. There have been attempts at plausible explanations such as the contraction of the hydrogen burning shell leading to a very concentrated release of energy which can only escape from the star if the overlying layers are expanded to reduce the effective opacity. A recent discussion argues that an increased interior luminosity does lead to the initial expansion and that, at a later stage, a change in state of the material produces an increased opacity which leads to accelerated expansion as the radiation tries to escape. There is in any case no reason to doubt the properties of the solutions of the equations of stellar structure which do predict that stars should expand and become red giants. The calculations predict that the star expands without any significant change in luminosity and this means that it moves rapidly to the right in the HR diagram (fig. 62).

Is there anything which will stop this contraction of the core and expansion of the surface layers? There are, at least, three ways in which the movement to the right in the HR diagram can be halted or slowed down. In the first case, if the central temperature of the star continues to rise, as it will according to the Virial Theorem if the central material remains a perfect gas, it may become high enough for the next nuclear fuel, helium, to be burnt through the reaction discussed in Chapter 4 (equation (4.27)). When this happens, the isothermal core is replaced by a helium burning convective core and, with the renewed release of nuclear energy, the release of gravitational energy and the central collapse are stopped. In the second case the central regions may become so dense that the pressure in the central regions is given by the degenerate gas law rather than the perfect gas law. At this stage further contraction becomes more difficult because of the high dependence on density of the pressure of a degenerate gas:

$$P_{\text{gas}} \simeq K_1 \rho^{5/3}. \tag{4.49}$$

Finally, if the surface temperature of the star becomes very low, there is once again a deep outer convection zone because of the existence of ionisation zones of hydrogen and helium just below the stellar surface. At about this stage the movement of the star to the right in the HR diagram is halted. This is because any further movement to the right would take the star into the Hayashi forbidden zone which was discussed in Chapter 5. The star does not cease to expand, but further increase in radius is now accompanied by an increase in luminosity rather than by a decrease in surface temperature ($L_s = \pi acr_s^2 T_e^4$). This is also shown in fig. 62.

The width of the main sequence

Before I continue the discussion of post-main-sequence evolution, I should comment on how the early evolution of stars affects a comparison of theoretical and observed main sequences for nearby stars. The observed main sequence, which is shown schematically in fig. 56, is quite broad. There are several reasons for this. We have seen in Chapter 5 that the position of the main sequence depends on stellar chemical composition, which is not identical for all nearby stars. In addition some stars may be binaries which cannot be resolved. If two stars have similar masses and surface temperatures, they can appear as a single star with up to twice the expected luminosity. Rapid rotation influences the total luminosity of a star but it also causes its radiation to be anisotropic. Calculations show that in whatever direction a rotating star is viewed it will appear to the right of the non-rotating main sequence. If none of these effects was present, the main sequence would be broadened somewhat by observational errors.

In addition to the above broadening of the *zero age main sequence*, stars remain close to the main sequence in their early post-main-sequence evolution and there is no obvious way of determining how long it is since a star first reached the main sequence. Any comparison of theory and observation should be between the zero age main sequences. Because both evolution and rotation place stars above the main sequence, the zero age observational main sequence must be towards the bottom of the observed band, with the stars below it being there because of observational errors or scatter in chemical composition.

There is one further factor which is important when a star cluster main sequence is studied. Not all stars in the cluster are at precisely the same distance and this leads to a further broadening of the observed main sequence, as individual distances to the stars are not known.

Dependence of early evolution on stellar mass – evolution of stars of high mass.

As mentioned earlier, the exact way in which a star evolves depends on its mass, and I now give a general description of the mass dependence. High mass stars have large convective cores when they are on the main sequence; the size of the convective core as a function of stellar mass has already been shown in Table 6 of Chapter 5. For such high mass stars, the supply of hydrogen which can be used in an initial burning phase is very large, because convection currents keep the entire convective core mixed. For very massive stars ($\gtrsim 10 M_\odot$) the mass in the convective region may grow as the star evolves, but if it does convection is not very efficient in the outer part of the core which is called a semi-convection zone. Even if the mass in the core decreases as the star evolves, a considerable fraction of the star's original hydrogen content may be burnt before the star exhausts its central hydrogen. When this does happen, the newly formed isothermal core may be almost immediately as large as the Schönberg–Chandrasekhar limit, so that the core collapses rapidly without an initial period of slow contraction.

Figure 63 shows how the hydrogen in the central regions of such a star is depleted as the star evolves. The hydrogen content X is plotted against fractional mass ($m \equiv M/M_s$) for several stages in the star's evolution. At each stage there is a convective core of uniform chemical composition, surrounded by an intermediate zone of variable chemical composition and an outer region with the original chemical composition of the star. The intermediate zone contains material which was in the convective core when the star was on the main sequence and its hydrogen content has been reduced by convection currents mixing it with material in which nuclear reactions have occurred.

During these phases of evolution the star does not move very far from the initial main sequence in the HR diagram, although it does show a slight increase in luminosity and decrease in surface temperature and this process leads to the appearance of a finite width in the observed main sequence as has already been mentioned. Thus from the surface properties of a star alone it is impossible to tell whether the star has just reached the main sequence or whether it has almost used

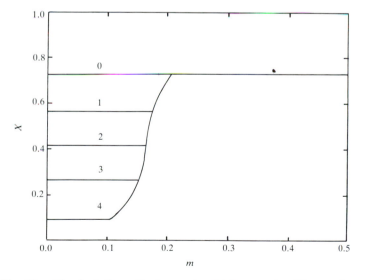

Fig. 63.   The depletion of hydrogen in a high mass star. The numbers denote successive stages in the star's evolution.

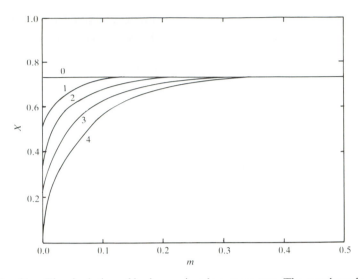

Fig. 64.   The depletion of hydrogen in a low mass star. The numbered curves refer to successive stages in the star's evolution. Although hydrogen is initially exhausted only in a very small central region, some hydrogen burning occurs further out than in a star of higher mass. This is caused by the relatively small temperature dependence of the rate of energy release by the PP chain.

up its central hydrogen and has moved some way from its initial position on the main sequence.

Once hydrogen is exhausted in the central regions, a nearly isothermal core of mass greater than the Schönberg–Chandrasekhar limit is soon produced and the star moves rapidly to the right in the HR diagram. What observational consequences can we expect from this rapid evolution? As I have already mentioned in Chapter 5, if stars travel through a particular region in the HR diagram in a very short time, we cannot expect to see many stars in that region at any given time. This immediately gives us a qualitative explanation of one of the properties of galactic cluster HR diagrams which I have already discussed in Chapter 2 (fig. 26) page 38. In fig. 26 it can be seen that in those clusters which possess main sequence stars of high luminosity and hence presumably high mass, there is a gap called the Hertzsprung gap between the main sequence and the red giants. Stars presumably cross that region very rapidly immediately after central hydrogen exhaustion and that is why very few stars (if any) are found in the Hertzsprung gap. The HR diagram for the nearby stars (fig. 56) shares the same property in that at high luminosity there is a distinct gap between the main sequence and the giant branch.

Evolution of low mass stars

I next consider low mass stars. These have only a small convective core or no convective core at all. This means that, when the hydrogen has been used up in the stellar core, it is only exhausted in a very small central region (fig. 64). As a result a small isothermal core is formed and it starts to contract long before its mass is comparable with the Schönberg–Chandrasekhar limit. It follows that the movement away from the main sequence is less rapid for low mass stars than for high mass stars and in addition the onset of degeneracy and a deep outer convection zone occur sooner for low mass stars, since they have higher central densities and lower surface temperatures than high mass stars before the central contraction and the outer expansion starts. In figs. 25 and 26 of Chapter 2, we have seen that there is no Hertzsprung gap in the HR diagrams of both globular clusters and old galactic clusters in which stars leaving the main sequence have low luminosity and hence presumably low mass. This is explained by the relatively slow evolution away from the main sequence for low mass stars. This will be discussed further below, when the significance of the phrase *old* galactic cluster will also become clearer.

The end of early evolution

Before presenting detailed results of calculations, it is perhaps necessary to define what I mean by early post-main-sequence evolution. The definition I adopt is not necessarily one of astronomical significance. I shall use the phrase early evolution to mean evolution from the main sequence until a stage is reached at which there is a real uncertainty in the theoretical calculations. We might, for example, reach a stage where, either because of lack of knowledge of the laws of physics or because of the inadequate size or speed of present computers, it is impossible to follow the evolution of the star any further.

One such specific uncertainty arises when the central regions of a star become degenerate, but not sufficiently degenerate to prevent the central temperature from continuing to rise to a value when the next set of energy releasing nuclear reactions occur. The onset of nuclear reactions in degenerate material can be very different from the onset of nuclear reactions in non-degenerate material. If nuclear burning starts in an ideal classical gas, this is potentially stable in the following sense. Suppose a small temperature rise occurs in such a region causing a large release of additional energy, because the rate of nuclear reactions depends on a high power of the temperature. If the stellar material is fairly opaque, this energy may not be able to escape as rapidly as it is produced. In that case a further local temperature rise will occur, but in an ideal classical gas this will also cause the pressure to increase and this will, in turn, lead to an expansion and cooling and a reduction in the rate of energy release. As an example of this process, I have already shown in Chapter 5 that, when stars approach the main sequence, the initial luminosity when nuclear reactions start is slightly higher than the luminosity when the star is on the main sequence.

If nuclear burning starts in a degenerate gas, the consequences can be quite different. The argument given above is still true as far as the further temperature rise. However, as I have discussed in Chapter 4, the pressure of a degenerate gas scarcely depends on temperature and this rise in temperature leads to a negligible pressure increase which is quite inadequate to cause expansion and cooling. In this case the local temperature rise, with a rapid increase in the release of energy and hence the rate of rise of temperature, continues until the temperature is high enough for the material to become non-degenerate. At this stage the material will expand, but, because of the runaway release of nuclear energy, the expansion is likely to be explosive. In some cases the observed explosions of stars may be due to the ignition of a nuclear fuel in a degenerate gas. It is not, however, clear that an explosion of the central regions of a star will necessarily lead to an explosion of the visible regions. For this to happen the central explosion must be intense enough to set the whole of the overlying layers of the star into violent motion.

In low mass stars, helium burning starts in material which is already degenerate and the ignition of helium in such circumstances is called the *helium flash*. The very high temperature dependence of the energy release from helium burning

$$\varepsilon_{3He} = \varepsilon_3 X_{He}^3 \rho^2 T^{40} \tag{4.28}$$

makes an explosive release of energy very likely. When such a helium flash occurs, it is very difficult to solve the equations of stellar structure accurately and no completely satisfactory study of the evolution of low mass stars past the onset of helium burning has yet been completed. Whereas in most stages of evolution nothing significant will happen to the structure of a star of one solar mass in a period of a million years, when the helium flash starts significant changes occur in the central regions in 100s. Even with a very large computer it is difficult to study the changes in the structure of a star in such rapid evolutionary phases. For such low mass stars, the onset of helium burning can be taken to mark the end of early post-main-sequence evolution. In more massive stars it appears that helium burning starts uneventfully and it is possible to study the evolution of the star until all of the central helium has been turned into carbon. If the central regions then become degenerate before the temperature is high enough for nuclear reactions burning carbon to occur, it is possible that an explosive energy release may occur at the onset of carbon burning.

I now discuss some detailed treatments of stellar evolution. In this chapter evolution at constant mass is assumed. The possible influence of mass loss is discussed in Chapter 7.

Detailed calculations of post-main-sequence evolution – Relatively massive stars ($3M_\odot \le M \le 10M_\odot$).

Calculations of the evolution of relatively massive stars by I. Iben are illustrated in fig. 65. In these calculations an initial chemical composition has been assumed for all of the stars which is similar to the chemical composition of population I stars in the Galaxy. In detail it is:

$$X = 0.708, \qquad Y = 0.272, \qquad Z = 0.020. \tag{6.2}$$

The reason behind this choice is that any relatively massive star which is in the phase of early post-main-sequence evolution must have been formed quite

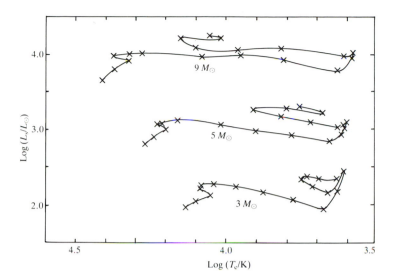

Fig. 65.   Post-main-sequence evolution of relatively massive stars.

recently in the galactic lifetime because its main sequence lifetime is quite short (see Table 5). It appears that population I stars are relatively young stars and that they have a higher content of heavy elements than the older population II stars. In the calculations it has been assumed that the stars evolve with constant mass and the best available values have been used for laws of opacity and energy generation. At the time that these calculations and the others to be described in this chapter were performed there was no clear evidence of important mass loss from massive main sequence stars. It is now known that this does occur and that it can have a significant influence on stellar evolution. This is discussed in the next chapter. It can be seen from fig. 65 that the evolutionary track of a single star is very complicated according to present theories.

R. Kippenhahn and his collaborators also studied the evolution of relatively massive stars and they followed them to a later stage in their evolution. They used a different chemical composition from that chosen by Iben:

$$X = 0.620, \qquad Y = 0.354, \qquad Z = 0.044. \tag{6.3}$$

The results obtained by Kippenhahn are shown in fig. 66. Where they can be compared with the results obtained by Iben, they are generally similar although there are some detailed differences. A comparison of the curves in figs. 65 and 66 gives an idea of what may be the uncertainties in the theory of the evolution of stars in this mass range. Kippenhahn's group has also considered the evolution of stars with different chemical compositions and they have found that quite small changes in the value of Z can lead to substantial changes in the evolutionary tracks. Their results for a different chemical composition

$$X = 0.739, \qquad Y = 0.240, \qquad Z = 0.021 \tag{6.4}$$

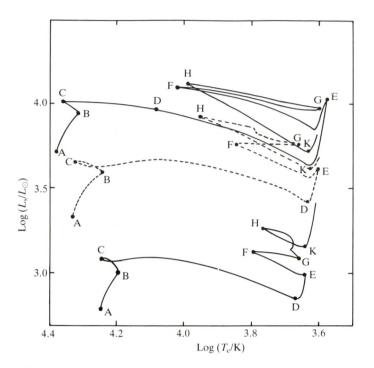

Fig. 66.   Post-main-sequence evolution of relatively massive stars. The stars have a different chemical composition from those of fig. 65 and are followed through a greater fraction of their life history. From the top the curves are for $9M_\odot$, $7M_\odot$ and $5M_\odot$.

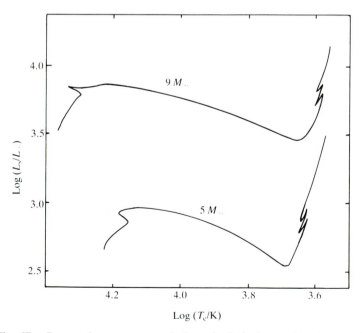

Fig. 67.   Post-main-sequence evolution of relatively massive stars, again with a different chemical composition.

Table 8. *Time (in 10^7 years) taken to reach lettered points on evolutionary tracks of fig. 66.*

$M_s/M_\odot$	B	C	D	E	F	G	H	K
5	5.37	5.62	5.91	5.94	6.76	7.04	7.83	7.86
7	2.56	2.60	2.65	2.66	3.15	3.31	3.56	3.57
9	1.59	1.65	1.66	1.67	1.86	1.94	1.99	1.96

are shown in fig 67. It can be seen that there are some substantial differences between these results and those of Iben.

Although the individual evolutionary tracks of stars are very complicated, according to current theories, this does not mean that these complications will all be reflected in the HR diagram of a star cluster. As has been mentioned earlier, some of the evolutionary phases are extremely rapid and this means that there are regions on the evolutionary tracks in which we are very unlikely to see stars. The times taken for different portions of the tracks of fig. 66 are shown in Table 8.

Evolution of a star of five solar masses

For one of the cases studied by Kippenhahn and his colleagues I will discuss the results in much greater detail. This is the star of five solar masses† with chemical composition given in (6.3). The evolutionary track of this star is shown in fig. 68 and the condition of the interior of the star at different stages in the evolution is shown in fig. 69. As the evolutionary track of the star is very complicated it is useful also to describe in some detail the state of the star at all of the points labelled in fig. 68. These are:

A The star is on the main sequence. It has a convective core containing about 21% of the mass of the star and nuclear reactions converting hydrogen into helium are essentially confined to about the inner 7% of its mass.

B The convective core has now shrunk to about half of its original size (in terms of mass contained) and a considerable fraction of the central hydrogen has now been consumed.

C The point of central hydrogen exhaustion. An isothermal core containing no hydrogen is formed and it very soon has a mass exceeding the Schönberg–Chandrasekhar limit, after which rapid collapse of the core occurs. A hydrogen burning region exists outside the isothermal core. This is initially quite thick but, subsequently, as it burns outward in the star, it becomes much narrower.

---

† It would be nice to refer this discussion to a particular star. However, as mentioned in Chapter 2, there is no giant star with a well-known mass. The two (main-sequence) components of the eclipsing binary U Ophiuci are observed to have masses of about $5M_\odot$ and this discussion can be regarded as a prediction of their future life history.

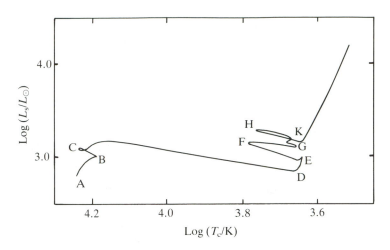

Fig. 68.   Post-main-sequence evolution of a star of five solar masses.

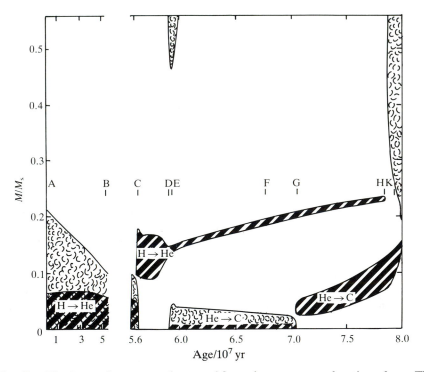

Fig. 69.   The internal structure of a star of five solar masses as a function of age. The heavily hatched areas are ones in which the named nuclear reactions are occurring. Convection zones are shown with a cloudy appearance.

From C to D the star moves very rapidly to the right in the HR diagram as the isothermal core, more massive than the Schönberg–Chandrasekhar limit, collapses rapidly and the outer regions expand. The central temperature is too low for helium to burn.

D At this point the star develops a deep outer convective region which at its maximum extent contains about 54% of the total mass of the star. The star has a structure appropriate to a largely convective star and its movement in the HR diagram becomes almost vertical just to the left of the Hayashi forbidden region, the luminosity and effective temperature are now following a track very similar to the one followed in the opposite direction as the star approached the main sequence (fig. 54), but the internal structure of the star is very different. The hydrogen shell source has become very thin and from here until point H it contains only about 1% of the mass of the star.

E The point of central helium ignition. Once the helium is burning, the star develops a new convective core containing about 5% of the mass of the star. Helium burns to carbon in the inner part of this convective core. Because the rate of helium burning depends on a very high power of the temperature (see (4.28)), significant energy release occurs only in about the inner 1% of the star.

F By this stage the convective core has shrunk somewhat and the central helium content has been considerably reduced. It is not immediately clear why the star reverses in direction in the HR diagram, although a reversal has already been found at B at a corresponding stage in core hydrogen burning.

G At this point the central helium content is reduced to zero and the star then has an almost isothermal core composed of carbon and the original admixture of heavier elements, which have not been affected by any of the nuclear reactions. From this stage onwards, helium burns to carbon in a shell outside the isothermal core.

H Ever since stage C, hydrogen has been burning in a shell which has been gradually moving further out in the star, as can be seen in fig. 69. The temperature of this hydrogen burning shell is determined partly by the properties of that part of the star inside it. At point H this temperature, which has been falling for some time becomes too low for further significant burning of hydrogen into helium and the hydrogen shell source does cease to exist. It may be noted from fig. 68 that, when the shell source does cease to be effective, the star abruptly changes its direction in the HR diagram.

K The star once again develops a deep outer convection zone, which this time contains more than 80% of the mass of the star. As the star becomes largely convective it once again moves more or less vertically in the HR diagram just to the left of the Hayashi forbidden zone. This convection zone penetrates into the region where all of the hydrogen has been converted into helium and mixes this material with outer layers which still contain their hydrogen. Thus the convection brings hydrogen from the outer part of the star down towards the region where helium is burning. If this hydrogen were brought to too high a temperature, it would ignite explosively. At the time of this work it was uncertain whether or not this would occur before:

L The ignition in the centre of the star of carbon, which burns through reactions such as

$$^{12}C + {}^{12}C \rightarrow {}^{24}Mg + \gamma, \tag{6.5}$$

$$^{12}C + {}^{12}C \rightarrow {}^{23}Na + p \tag{6.6}$$

and

$$^{12}C + {}^{12}C \rightarrow {}^{20}Ne + {}^{4}He. \tag{6.7}$$

From later calculations It seems that carbon burning probably starts before hydrogen is re-ignited and this presumably leads to the formation of a carbon burning convective core. Although at this stage the material in the centre of the star has become a slightly degenerate gas, it does not appear that the carbon ignition is explosive.

These results have been described in some detail in order to show what stage studies of stellar evolution have now reached. The evolutionary track of a star appears to be very complicated with multiple passages across the HR diagram. From this and other calculations it appears that each change of direction in the diagram is associated with a change in the importance of some energy source. It should be stressed once again that the time taken in the advanced evolutionary phases is very small compared to the duration of the early phases, so that we do not expect to observe many stars in an advanced stage of evolution and we do not expect all of the complications of an individual evolutionary track to show up in a star cluster HR diagram. A further effect which shortens late stages of stellar evolution will be discussed in Chapter 9. As I shall also explain in Chapter 9 mass loss from a $5M_\odot$ star must modify the evolution which has just been described

The ages of young star clusters

I now turn to the comparison of theory and observation. I have already mentioned in Chapter 3 that we do not know enough about the properties of individual stars to make it worth while to try to discuss their properties in detail, but we can hope to explain why the HR diagrams of galactic and globular clusters have the shapes which they are observed to have. In particular, we can hope to determine the approximate ages of star clusters, that is how long it is since the stars in the clusters were formed. The shapes of the evolutionary tracks of relatively massive stars just after the main sequence stage can be used to estimate the ages of

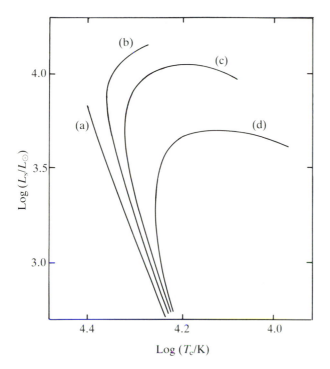

Fig. 70. Isochrones for young clusters. (a) is the main sequence and (b), (c), and (d) are are loci of stars of ages 10^7, 1.66×10^7 and 2.65×10^7 years. The Hertzsprung gap does not appear in this diagram because I have not attempted to show the relative density of stars at different points in the loci.

young and moderately old star clusters; these are clusters in which some stars in the mass range I have been discussing have not evolved too far away from the main sequence. From the main sequence lifetimes shown in Table 5 of Chapter 5, this means that the clusters should be no more than a few hundred million years old.

Let me consider first an idealised situation in which I have a cluster of stars of identical chemical composition all of which reach the main sequence at precisely the same time. I am thus considering a group of stars which differ only in mass. As time passes, the stars will evolve away from the main sequence, and because of the dependence of luminosity on mass, the more massive stars will evolve more rapidly. If I plot the evolution of stars of different mass in an HR diagram, I can also insert in the diagram lines of equal time or *isochrones*. Thus, for example, I can mark on each evolutionary track the position of the star after 10^8 years. These points can then be joined up to produce the 10^8 year isochrone. I have already drawn such an isochrone schematically in fig. 30 of Chapter 2 and some more are shown in fig. 70. If all the stars in the cluster had the same age and chemical composition, they should lie on a single isochrone in the HR diagram and the age of the cluster could be deduced.

Table 9. *Approximate ages (in years) of young and intermediate age galactic clusters*

h and χ Persei	10^7
Pleiades	7×10^7
Praesepe, Hyades	4×10^8
NGC 752	10^9

In practice the problem is rather more difficult. To begin with the theoretical calculations give L_s and T_e while the observations give V and $B–V$. In Chapter 2 I have discussed the relations between L_s and V and T_e and $B–V$ which are needed before a comparison between theory and observation can be made. There are uncertainties in this conversion from theoretical to observed qualities which must not be forgotten. Secondly I cannot expect all stars in a cluster to reach the main sequence at the same time. Stars may not all have been formed at the same time, but even if they were, the time taken to reach the main sequence varies with the mass of the star. If the cluster is old enough, the spread in arrival time at the main sequence will be small compared to the age of the cluster, but for young clusters the isochrones must be drawn after allowance has been made for the time stars take to reach the main sequence. Although cluster HR diagrams are rather sharply defined, they are not completely sharp and all the stars certainly do not lie on one theoretical isochrone. However, it is possible to find the isochrone which agrees best with the cluster HR diagram and, using this method, the ages of various galactic clusters have been estimated and the results are shown in Table 9.

It is clear that the simplest idea that all stars in a cluster have exactly the same age and chemical composition is not correct. Although inaccuracies in the observations, variations in chemical composition between the stars in a cluster and the effects of rotation on the properties of some stars all affect this comparison between theory and observation, it seems that the major cause of the spread in the HR diagram of young clusters is that there is a finite period of star formation in a cluster. To obtain agreement between theory and observation, it must be assumed that star formation in any one cluster may continue for up to a few tens of millions of years. This is particularly important in the youngest clusters whose age is comparable with this and in some of these star formation is probably still proceeding.

As well as simply comparing the shapes of theoretical and observational HR diagrams, there are other more detailed comparisons between theory and observation which can be made and two of these will be discussed below.

The initial mass function and the relative numbers of red giant and main sequence stars

If I suppose once again that the only important factor differentiating stars in a cluster is mass, I can describe the cluster by what is known as its *initial mass function*. Suppose the number, dN, of stars formed in the cluster with masses between M and $M + dM$ is given by:

$$dN = f(M)\,dM. \tag{6.8}$$

Then $f(M)$ is called the initial mass function for the cluster. It is called the *initial* mass function because it may be altered as stars evolve if they suffer serious losses of mass. The initial mass function is determined by the process of star formation. As has been mentioned in Chapter 5, there is as yet no reliable theory of star formation and this means that there is no initial mass function predicted by theory. It is possible to try to obtain some information about the initial mass function from observations and it is of particular interest to try to obtain the initial mass function for different systems of stars to see whether it is always approximately the same. If it were always the same, it would suggest that there was some process which always made a gas cloud fragment in the same way.

It is not easy to obtain the mass function of any group of stars as their masses cannot usually be measured directly and have to be estimated from comparison between theory and observation. E Salpeter found, for the stars in the solar neighbourhood:

$$f(M) = CM^{-2.33}, \tag{6.9}$$

where C is a constant. It is now believed that there are important deviations from this law for massive stars and it remains uncertain how many very low mass ($\lesssim 0.1 M_\odot$) stars there are, even very close to the Sun. Although there is still no very clear evidence about how the initial mass function varies with position inside our own Galaxy and from galaxy to galaxy, it does appear that the star formation in some exceptional galaxies, starburst galaxies, is strongly biased towards massive stars.

When we look at the HR diagram of a cluster, the stars that we see away from the main sequence do not vary very much in mass. This is because the speed of post-main-sequence evolution, once it starts, depends on a relatively high power of the mass of the star. To put it another way, away from the main sequence there is a great similarity between an isochrone and the evolutionary track of a single star. This is illustrated in fig. 71. The evolutionary tracks of stars of different masses are shown and an isochrone is superimposed and this only differs significantly from the evolutionary track of the most massive star near to the main sequence. Provided the initial mass fuction is not an extremely steeply varying function of the mass of the star, this means that the density of stars in any region of cluster HR diagram away from the main sequence should be directly proportional to the time that an individual star spends in that region.

I have already used this argument in a weak sense to explain why I do not expect to see many stars in the Hertzsprung gap region of young galactic clusters. I can also compare the number of red giants with the number of stars near to, but above, the main sequence. This is illustrated for the young double cluster of h and χ Persei in fig. 72. Here the observed ratio of stars in regions B and C, as well as their relation to the number on the main sequence at A, can be compared with the ratio

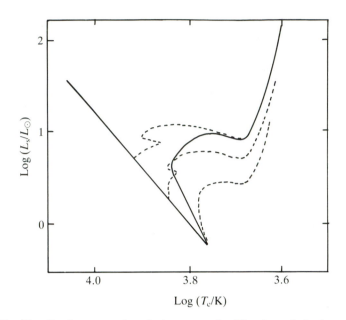

Fig. 71.   Isochrones and evolutionary tracks. The three dashed curves are evolutionary tracks for stars of three different masses. The solid curve is an isochrone.

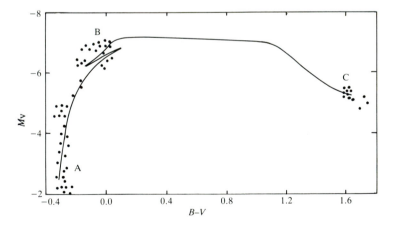

Fig. 72.   The HR diagram of the double cluster h and χ Persei on which is superimposed the evolutionary track of a star of $15.6 M_\odot$.

predicted by theory. At present the agreement between theory and observation is not entirely satisfactory and this may lead to some modifications in the theory. One problem that affects all comparisons between theory and observation is that there may be important mass loss as stars evolve. If massive stars lose a large amount of mass when they are close to the main sequence, they do not move so far to the right in the HR diagram and this affects the relative numbers of red and blue supergiants in a cluster like *h* and *χ* Persei.

Cepheid variable stars in galactic clusters

I have mentioned several times that there are very few stars in the Hertzsprung gap in galactic clusters. There are a few cepheid variables in galactic clusters. We can see from fig. 28 of Chapter 2 that they lie between the main sequence and the giant branch and they *are* situated in the Hertzsprung gap. As these variables occur in a well-defined region of the HR diagram, it is of importance to ask why stars in that region should be variable while other stars in the cluster are not.

The cepheid variables have a luminosity which varies in a regular, but not smooth manner. Some typical cepheid light curves are shown in fig. 73. The characteristic shape of the curve varies with the period of the oscillation. When cepheid variables, which are named after the star *δ* Cephei which has been observed since 1784, were first discovered, it was thought that the light variations might be due to eclipses in a binary system, although the shapes of the curves were unusual for binaries. Subsequently study of the spectra of the stars showed that

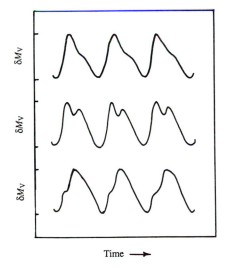

Fig. 73.   Light curves of cepheid variables. The characteristic shape of the light curve changes with period of oscillation. In the figure the top curve corresponds to the cepheid of shortest period.

they were undergoing radial pulsations; that is, the radius of the star was varying as well as its light output.

Initially it was thought that the stars were varying because they had received some external disturbance such as might be caused by a close passage of another star. Just as a simple pendulum has a characteristic period of oscillation which is independent of the manner in which it is set oscillating, a star has a characteristic period of radial oscillations. This was calculated and was found to be comparable with the observed periods of variable stars. A simple pendulum does not oscillate for ever, as it is gradually damped by air resistance and friction at the support. Similarly internal viscosity and other damping processes affect the oscillations of a star and it was soon realized that oscillations would be damped at such a high rate that it was unlikely that the variable stars arose accidentally. This view was strengthened when it was discovered that the variables occupied a compact region in the HR diagram, It seemed much more likely that the oscillation was stimulated by some process within the star and that this process was operative only for stars in a particular region of the HR diagram.

Recent calculations have given strong support to this view. No stars are in a perfectly steady state: in many cases small departures from the steady state are damped as rapidly as they arise, but in others the fluctuation may be amplified. In the case of stars situated in the region of the Hertzsprung gap, it has been found that small radial oscillations will increase in amplitude and calculations of the growth of these oscillations have gone a long way towards explaining detailed properties of the variable stars such as the shapes of the light and velocity curves. There are still some detailed disagreements between theory and observation, but there seems no doubt that many stars situated in the Hertzsprung gap should be variable stars. The final steady-state oscillation of a cepheid variable arises as follows. If the star experiences a small accidental oscillation, this is initially amplified and we say that the star is unstable to small disturbances. As the amplitude of the oscillation grows, the damping forces which we have mentioned earlier also increase and finally a steady oscillation results in which the amplifying and damping processes are just in balance. For the vast majority of main sequence stars the initial small disturbances do not grow and the stars are not variable.

The early evolution of low mass stars

I have mentioned earlier in this chapter two ways in which the evolution of low mass stars differs from that of massive stars. In the first place, when all of the hydrogen has been burnt in the centre of the star, the resulting isothermal core has a mass which is smaller than the Schönberg–Chandrasekhar limit. As a result the star does not move so rapidly to the right in the HR diagram. Secondly, the stars become degenerate in the centre before the onset of helium burning and because of this what we define to be early evolution ends at the start of helium burning. Calculations by Iben of the evolution of low mass stars are shown in fig. 74. As in his studies of relatively massive stars, these calculations are for stars with

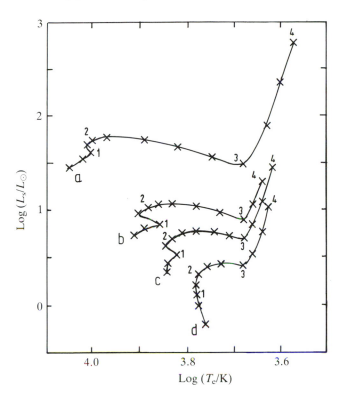

Fig. 74. Post-main-sequence evolution of low mass stars. Curves (a), (b), (c), (d) are for stars of $2.25M_\odot$, $1.5M_\odot$, $1.25M_\odot$ and $M_\odot$ respectively.

a population I chemical composition. Thus they are directly relevant to the stars in old galactic clusters rather than the globular clusters whose stars have a much lower abundance of the heavy elements. The times taken to the different points marked on the theoretical curves are shown in Table 10.

Isochrones obtained from the evolutionary tracks of these stars can be used to obtain theoretical HR diagrams for old galactic clusters such as M67 and NGC188. These agree with the observational HR diagrams in having no significant Hertzsprung gap (see fig. 75). As in the case of the young galactic clusters, a cluster age can be estimated by comparison of theoretical and observational HR diagrams and recent estimates of ages for these old galactic clusters are shown in Table 11. Other workers have calculated evolutionary tracks for stars in this mass range with much lower heavy element content and these lead to estimates of the ages of globular clusters. These estimates are also shown in Table 11. The estimated ages of the old galactic clusters are lower than they were in the first edition of this book and the ages are still uncertain by at least $\pm 10^9$ yr. The estimated ages of globular clusters derived by different authors vary by at least $\pm 3 \times 10^9$ yr around the figure given in Table 11 and there may also be a spread of ages of about 3×10^9 yr between the oldest and youngest globular clusters.

Table 10. *Time (in 10^9 years) taken to reach numbered points on evolutionary tracks of fig.* 74

$M_s/M_\odot$	1	2	3	4
1.0	6.71	9.20	10.35	10.88
1.25	2.83	3.55	4.21	4.53
1.5	1.57	1.83	2.11	2.26
2.25	0.48	0.52	0.55	0.59

Table 11. *Estimated ages of two old galactic clusters (M67, NGC188) and globular clusters. The ages are in 10^9 years.*

M67	NGC188	Globular clusters
5	7	15

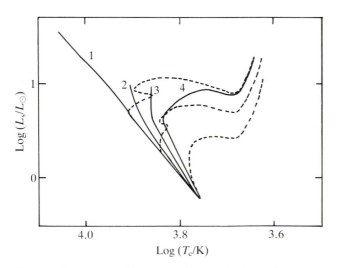

Fig. 75.   Isochrones for old clusters. Curve 1 is the main sequence and curves 2, 3 and 4 are loci of ages 0.5×10^9, 1.5×10^9 and 2.25×10^9 years.

In the case of low mass stars there is another uncertainty in their properties which I have mentioned briefly in Chapter 3. The lack of a good theory of convection was there summed up in the uncertain value of the mixing length parameter, α. As I mentioned earlier, the value of α is frequently chosen to provide a good solar model and it is then used in other stars. It is however quite possible that doubts about the value of α could lead to uncertainties of several hundred degrees in the effective temperatures of low mass main sequence stars.

The evolution and age of the Sun

Included amongst the stars of the mass range and chemical composition studied by Iben is the Sun. Although the Sun is in the main sequence region, it must have evolved somewhat from its initial main sequence position. Naturally, as the subject of stellar structure and evolution has developed, a considerable effort has been devoted to trying to account for all of the observable properties of the Sun, about which we know so much more than we know about any other star. Ideally, if we knew the chemical composition of the Sun in fine detail and all of the physical laws accurately, we should calculate its initial main sequence position and then follow its evolution until its actual position is reached and this would then give its age.

In fact there are many difficulties in this procedure. These are not really any worse for the Sun than for any other star except that we can make really accurate measurements of so many properties of the Sun, mass, radius, luminosity, surface temperature, etc., and we expect to find a theory which agrees in detail with all of these. There are uncertainties in the solar chemical composition, the laws of opacity and energy generation, the theory of convection and the conversion from L_s, T_e to V, $B-V$. In addition, we have some independent information about the age of the Sun. Geological evidence from the radioactive elements in the Earth's crust shows that the Earth must have been solid for the last 4.5×10^9 years and this makes it almost certain that the Sun must have an age greater than this. Attempts are therefore made to fit all of the observed properties of the Sun as well as can be done with a solar age a little greater than 4.5×10^9 years and in general it seems that the immediate post-main-sequence evolution of the Sun is reasonably well described.

Neutrinos from the Sun

There remains one question mark over the properties of the Sun and other low mass stars. In Chapter 4 I described how an attempt has been made to detect the neutrinos which are emitted by hydrogen burning reactions near the centre of the Sun. I did not mention there that the higher the energy of a neutrino, the greater is the likelihood that it can be detected because the probability of detection is generally proportional to the square of the neutrino energy. The neutrino emitted by the β-decay of ^8B in reactions (4.15) is much more energetic than the other neutrinos emitted in hydrogen burning and according to theory, the number of reactions going through that branch of the PP chain at the estimated central temperature of the Sun is proportional to T^{18}. As a result of this high dependence on temperature, if these neutrinos could be detected, the central temperature of the Sun would be determined with a high degree of accuracy and it could be compared with the central temperature predicted by theoretical calculations.

In the early 1960s estimates suggested that these neutrinos *could* be detected with great difficulty and an elaborate experiment was mounted to try to do this. Neutrinos do not readily advertise their presence even if they do interact with matter. The only way to detect them is to cause a stable atomic nucleus to be transformed into an unstable nucleus by capturing a neutrino and then to observe the β-decay of the unstable nucleus. If the experiment is to be successful, the neutrinos must be captured in a place where there are no other particles which will produce the same unstable nucleus. For this reason the experiment was placed in a deep mine to shield it from the effect of cosmic rays. As mentioned in Chapter 4, the nucleus chosen was ^{37}Cl, one of the stable isotopes of chlorine which accounts for about one-quarter of natural chlorine. The process to be studied is then:

$$^{37}\text{Cl} + \nu_e \rightarrow {}^{37}\text{Ar} + e^-, \tag{4.19}$$

followed later by:

$$^{37}\text{Ar} \rightarrow {}^{37}\text{Cl} + e^+ + \nu_e. \tag{4.20}$$

The chlorine was in perchloroethylene (cleaning fluid), C_2Cl_4 and 400 000 litres were used. The argon which is produced has to be removed from the tank before it β-decays. It was found, in a pilot experiment, that if another inert gas, helium, was bubbled through the tank the argon could be extracted with the helium and its decay subsequently observed.

Theoretical calculations predicted the rate at which the neutrinos would be captured. Because the mass of ^{37}Ar is greater than that of ^{37}Cl, the bulk of the neutrinos emitted in the PP chain, which come from the conversion of two protons to deuterium, are not energetic enough to be captured by ^{37}Cl. Almost all of predicted captures involved the high energy ^{8}B neutrinos. The number of neutrinos detected was lower than the theoretical expectation. This did not initially cause great alarm. The implication was that the central temperature of the Sun was slightly lower than was thought. It was expected that a new theoretical model of the Sun could be constructed, taking account of uncertainties in such things as the opacity law, the theory of convection and the solar chemical composition, which would have the required lower central temperature. This has never been achieved, the main problem being the difficulty in reducing the value of T_c without changing L_s and T_e. There continues to be a factor of about three between the number of neutrino captures predicted by theory and the actual number of captures.

Currently the hope is that the cause of the discrepancy between theory and experiment lies in the properties of the neutrino rather than in the physical conditions in stellar interiors. There are now known to be three types of neutrino, the electron neutrino, ν_e, the muon neutrino, ν_μ, and the tauon neutrino, ν_τ. Of these, only ν_e can produce the reaction (4.19). Particle physicists think that it is possible that these three types of neutrino are not totally distinct and that they can turn into one another. In that case a beam of ν_e emitted from the central regions of the Sun could be a mixture of ν_e, ν_μ and ν_τ by the time it reaches the experiment. This might resolve the problem and allow the solar T_c to have its theoretical value.

There are now two further experiments which use the reaction

$$^{71}\text{Ga} + \nu_e \rightarrow {}^{71}\text{Ge} + e^-. \tag{6.10}$$

They have the names SAGE and GALLEX. These experiments took a long time to set up because gallium is a much rarer element than chlorine and a large quantity of it is required. The advantage of gallium as a target is that it is sensitive to the low energy neutrinos which are emitted in the first reaction of the PP chain. Because almost all the reactions are believed to go through the PPI chain, the number of these low energy neutrinos is scarcely affected by the value of T_c. Thus, if the experiments do not detect the expected number of neutrinos, the disagreement cannot be resolved by a change in T_c. Present indications are that there is a disagreement, and this may be most easily resolved if there is a mixing of neutrino types. It is to be hoped that this matter will be resolved within a few years.

Helioseismology

There is one further way in which information is being obtained about the solar interior and there is also the hope that the technique may be able to be extended to the study of some other stars. The solar surface exhibits some low level oscillations in its properties which were given the name five-minute oscillations when they were first discovered, because that was their characteristic period. The Sun like any dynamical system will possess a spectrum of normal modes of oscillation, whose periods and other properties will depend on the internal structure of the Sun including such properties as its density and chemical composition and also on the variation of its rotation velocity with depth and its internal magnetic field. For any theoretical model of the Sun, it is possible to determine the properties of its oscillation modes, most of which will not be spherically symmetrical and will involve a distortion of the shape of the Sun.

The observations can be analysed into a discrete set of frequencies and it can then be asked whether the observed frequencies match those of the theoretical model. Helioseismologists tend to find that the present standard model of the Sun, which as we have seen does not provide the observed neutrino flux, is in reasonable agreement with observations. In contrast, some non-standard models which have been proposed to solve the neutrino problem, do not pass the helioseismology test. The observations are some of the most difficult in astronomy, but it would seem that they have a very promising future.

The evolution of very low mass stars

In Chapter 5 I have stated that below a mass of the order of $0.1M_\odot$ stars do not even have a main sequence hydrogen burning phase. Such low mass stars, which are perhaps more like planets than stars, have central temperatures which reach a maximum value which is too low for significant nuclear burning and which then decrease. As the central temperature of the star falls, so do both its luminosity and radius and the further evolution of such a star is shown schematically in fig. 76. As the star reaches the bottom of the evolutionary track shown in fig. 76 it is becoming cold and very dense and its properties resemble those of white dwarfs, except that it has a lower surface temperature and luminosity than the white dwarfs which we observe. These very low mass stars thus evolve directly to the white (or more realistically black) dwarf state. The white dwarfs which are observed are mainly much more massive than this and they have had a more complicated life history. While extremely low mass stars do not have a normal main sequence lifetime, slightly more massive stars do ignite hydrogen in their interiors. For a range of masses above $0.1M_\odot$, hydrogen burns in the central regions of the star but once the central hydrogen is exhausted the temperature never becomes high enough for helium burning to start. I have already stated in Chapter 5 that helium will not burn in a pure helium star if the mass is less than about $0.35M_\odot$; for a star which is initially of normal chemical composition the critical mass is somewhat larger. A schematic evolutionary track of a star of mass between $0.1M_\odot$ and $0.4M_\odot$, which burns its central hydrogen but not its central helium, is also shown in fig. 76. It can be seen that the initial post-main-sequence evolution is similar to that of rather more massive stars shown in fig 74 but the luminosity reaches a maximum and then declines without helium burning starting.

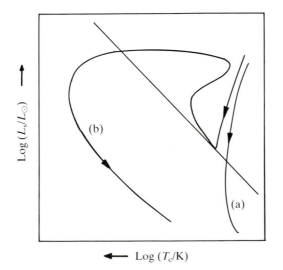

Fig. 76.   The evolution of stars of very low mass. (a) refers to a star of lower mass than $0.1\ M_\odot$ and (b) refers to a star of mass between $0.1M_\odot$ and $0.4M_\odot$.

It is not, in fact, believed that stars in this mass range will have completed their main sequence evolution during the lifetime of the Galaxy, because of the dependence of main sequence lifetime on stellar mass. This may therefore be only a prediction of behaviour to occur later in the history of the Universe.

This implies that the stars which are currently white dwarfs cannot have reached that state by evolution at constant mass. White dwarfs are observed in galactic clusters with ages of order 10^8 years and any star which can have evolved in such a time must originally have been significantly more massive than the Sun. The manner in which stars become white dwarfs will be discussed further in Chapters 9 and 10.

Summary of Chapter 6

The early post-main-sequence evolution of a star depends on its mass. A star remains near the main sequence until most of its central hydrogen has been converted into helium. For high mass stars, with large convective cores, more hydrogen can be burnt before this state is reached than in low mass stars without convective cores. When the central regions contain no further hydrogen they become almost isothermal. When an isothermal core contains more than about 10% of a star's mass, the central regions contract rapidly and the outside simultaneously expands. This happens almost as soon as the central hydrogen is exhausted in high mass stars and they move rapidly into the red giant region. This accounts for the Hertzsprung gap in some galactic clusters. Low mass stars also become red giants, but less rapidly, and this agrees with the absence of a Hertzsprung gap in globular clusters and the remaining galactic clusters. The calculated evolutionary tracks of stars of different masses, combined with the assumption that stars in a cluster differ mainly in mass, give an estimate of cluster age. Ages found vary from a few million years for young galactic clusters to more than 10^{10} years for globular clusters.

The straightforward calculation of early stellar evolution ends if the central regions become degenerate while the central temperature is still rising. Nuclear reactions in a degenerate gas may occur explosively, thus resembling a nuclear bomb rather than a nuclear reactor. Because of this, it has proved difficult to study the evolution of low mass stars beyond the start of helium burning. For massive stars, the material remains a perfect gas after helium has burnt. The calculated evolutionary tracks of high mass stars are very complicated with several crossings of the HR diagram. Observations are unlikely to show the full complexity of these tracks because some of the stages of evolution are very rapid and we cannot expect to observe many stars in these stages.

The early evolution of and, particularly, the present state of the Sun has been much studied. An attempt to verify conditions in the centre of the Sun by detecting neutrinos emitted in nuclear reactions has given an uncertain result. The number of neutrinos detected has been fewer than expected. Initially it was thought that this implied that the solar central temperature was lower than calculations predicted. Now it seems rather more probable that an unforeseen neutrino property is causing the disagreement between theory and experiment.

7

Mass loss from stars

Introduction

In most of my previous discussion I have assumed that stars spend their entire life with a constant mass, whose chemical composition changes as a result of nuclear reactions. I have also mentioned that, in fact, mass loss from stars is important and I shall now say more about this subject. Observationally mass loss is apparent in the explosions of supernovae and, to a lesser extent, novae and it is also clear that planetary nebulae are formed of mass ejected by stars. Evidence of mass loss from ordinary stars has only been well-established since the development of space astronomy as will become clear in what follows. In this chapter I shall restrict myself to a discussion of mass loss from single stars or from binary stars whose separation is so great that the two components evolve independently. In the next chapter I shall discuss mass exchange between components in close binary systems, which may also involve mass loss from the entire system.

The solar wind

It has been known since about 1960 that the Sun is losing mass at a rate of between one part in 10^{14} and one part in 10^{13} a year. This loss is known as the *solar wind* because it flows through interplanetary space and past the Earth with a velocity of several hundred kilometres a second. This is too low a rate of mass loss to be observed by direct visual observations of the Sun, but the solar wind particles have been detected by space probes and the discovery of the solar wind was one of the first astronomical measurements made by the space programme. Before the solar wind was discovered, its possible existence was suggested in two different ways. The behaviour of the tails of comets, when they were close to the Sun, could be understood if they were continuously bombarded by a stream of electrically charged particles emitted by the Sun. In addition, some theoretical discussions of the structure of the high solar atmosphere suggested that it could not be in hydrostatic equilibrium and that mass loss must occur. The solar wind mass loss is unimportant as far as solar evolution is concerned; it is a direct loss of mass comparable with the mass equivalent of solar radiation. If other stars only lost mass at a similar or rather greater rate, this would be both unobservable and of no evolutionary interest. In fact, much greater rates of mass loss are observed from a wide variety of stars.

Let us consider first why it is that the Sun loses mass. It has been known for over a century that the outer atmosphere of the Sun, beyond the visible photosphere with its temperature of $\simeq$ 5800K, does not fade quietly into the interstellar medium. Above the photosphere, the temperature passes through a minimum of $\simeq$ 4300K and then increases through the *chromosphere* and rapidly through the transition region to the *corona*, where the temperature exceeds 10^6K. This is not a true thermodynamic temperature. Conditions are very far from thermodynamic equilibrium and the radiation field is totally different from the Planck form. The density of the corona is very low and the temperature is a kinetic temperature related to the average kinetic energy of the particles. For some reason that needs to be understood, the small fraction of the particles in the solar atmosphere that forms the corona has far more than its share of the available energy. If it is simply accepted that the corona does have such a high temperature, it is possible to show that it cannot sit in static equilibrium on top of the Sun, unless it is surrounded by interplanetary and interstellar matter with a pressure high enough to hold it in. Such matter does not exist and the corona continually escapes to form the solar wind.

It is not difficult to demonstrate that a high temperature static corona is not possible. Suppose that the corona is in both thermal and hydrostatic equilibrium. In the corona any transport of energy will be by thermal conduction and the conductive luminosity must be the same at all radii. If the coefficient of heat conduction is λ, the conductive luminosity at any radius r will be

$$L_{\text{cond}} = -4\pi r^2 \lambda \, dT/dr. \tag{7.1}$$

If this does not vary with radius,

$$\frac{d}{dr}\left(r^2\lambda \frac{dT}{dr}\right) = 0. \tag{7.2}$$

At the temperature observed in the corona, the coefficient of heat conduction appropriate to an ionised gas is known to be

$$\lambda = \lambda_0 T^{5/2}, \tag{7.3}$$

where λ_0 is a constant. If (7.3) is inserted into (7.2), this integrates to give

$$T/T_0 = (r_0/r)^{2/7}, \tag{7.4}$$

where T_0 and r_0 are constants which can be taken as any radius in the observed corona and the corresponding temperature.

It is now necessary to solve the equation of hydrostatic equilibrium which can be written

$$\frac{dP}{dr} = -\frac{GM_\odot\rho}{r^2}, \tag{7.5}$$

if the mass in the corona is a very small fraction of the mass of the Sun. Use of the equation $P = \mathcal{R}\rho T/\mu$ and (7.4), together with the assumption that μ does not vary with radius then gives

$$\frac{d\rho}{dr} = \frac{2\rho}{7r} - \frac{GM_\odot\mu\rho}{\mathcal{R}T_0 r_0^{2/7} r^{12/7}}. \tag{7.6}$$

At large values of r the second term on the right hand side of (7.6) must become small compared to the first term and, if the second term is neglected, the equation integrates to give

$$\rho = kr^{2/7}, \tag{7.7}$$

where k is a constant. Combination of (7.4) and (7.7) then shows that P approaches a constant value at large radii and it is this constant value which cannot be balanced by the known interplanetary and interstellar pressure. As a result the corona must expand into interplanetary space.

It is easy to show that the corona cannot be static, but it is more difficult to discuss the dynamics of the solar wind and I shall not attempt to do so. However, once the relationship between the corona and the solar wind is established, it is necessary to ask why the Sun has a corona and what possible clues that might give to the existence of coronae on and of mass loss from other stars. The study of the solar corona has in the past been very difficult because it has such a low luminosity and a high temperature. It emits little radiation in the visible region of the spectrum and this radiation can only be observed during a solar eclipse or through artificial eclipses in a device known as the coronagraph. The development of space astronomy means that X-rays and ultraviolet radiation from the solar corona can be observed at other times.

The solar convection zone

The Sun has a region just below its visible surface where theoretical calculations predict that the major energy transport is by convection; the occurrence of outer convection zones in low mass stars has been mentioned in Chapter 5. This belief is strengthened by the observed appearance of the surface of the Sun. It is covered with a cellular pattern of the type that has been mentioned in connection with convection in Chapter 3, with rising and falling elements of matter. It used to be thought that the convection zone was directly responsible for the existence of the corona. The suggestion was that the convective motions caused sound waves to propagate into the outermost layers of the Sun and that, as the waves moved outwards into a region of decreasing density, they steepened and became what are known as shock waves, which then dissipated their energy and heated the corona. The present view is that the existence of the corona and solar wind is more closely related to the magnetic field in the outer layers of the Sun, although the properties of the magnetic field are also believed to be controlled by the convection zone. The corona may be heated by magnetohydrodynamic waves or by events similar to the visible solar flares.

The surface of the Sun is very complex and the approximation of spherical symmetry, which I have used to discuss the overall properties of stars and in my discussion of a static corona, is not applicable. In addition to the granular appearance produced by convection, there is the sunspot cycle. *Sunspots* are regions of the surface which are cooler than their surroundings and which contain intense magnetic fields. There are also *solar flares* in which a large amount of energy is emitted from a small region in a very short time and *prominences* in which material is raised above the solar surface in a large arc. All of these phenomena are influenced if not controlled by the solar magnetic field. Some of the magnetic field lines which leave the solar surface form closed loops and return to the surface again, but others extend far into interplanetary space and are effectively open field lines. The hottest part of the solar corona lies along the closed loops and the solar wind flows along the open field lines.

The magnetic field is produced by electric currents which flow in the solar atmosphere and these currents, and hence the magnetic field, would decay in a time very short compared to the present lifetime of the Sun, if they were not regenerated. It is believed that the convective motions are responsible for maintaining the magnetic field through what is known as a dynamo process. It is therefore to be expected that other stars with cool surfaces, which have outer convection zones, are also likely to have magnetic fields, coronae and mass loss. The loss of mass will in turn be easier if a star with a deep outer convection zone is also a red giant. The reason for this is that the velocity matter has to acquire in order to escape from a star is $v_{esc} \equiv (2GM_s/r_s)^{1/2}$, which for a star of a given mass decreases as the radius is increased. I have explained in the last chapter that stars do have deep convection zones when they become red giants.

Although these purely theoretical ideas are suggestive, direct observations of the properties of the very outer layers of cool stars only became possible with the development of telescopes carried by satellites, which can study the ultraviolet region of their spectrum. The temperatures of the solar upper chromosphere, transition region and corona are such that their principal radiation is in the ultraviolet or even in the X-ray region of the spectrum. This radiation cannot be observed by telescopes placed on the Earth's surface because it is strongly absorbed by the Earth's atmosphere. This difficulty is resolved by the placing of telescopes into orbit although the earliest results were obtained from rockets. In particular, knowledge of the Universe at ultraviolet wavelengths has been transformed by observations with the International Ultraviolet Explorer (IUE) satellite, which was put into orbit in 1978 and which is still operational at the time of writing (1993). Studies with IUE have cofirmed that coronae and chromospheres and also mass loss are common in late type stars with outer convection zones. It has also become clear that some stars have very intense flares and that others have star spots much more dramatic than sunspots. In some cases the rate of mass loss is clearly many orders of magnitude greater than that associated with

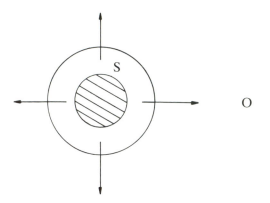

Fig. 77.   Mass flows outwards as shown from the surface S of a star. Because the material being lost is at least partially transparent to radiation, an observer at O sees very wide spectral lines from material in both hemispheres of the star.

the solar wind and this means that it can influence the evolution and the final mass of the star. For example there may be significant mass loss from low mass stars as they climb the giant branch towards the onset of helium burning.

Mass loss from stars is detected by the observation of very wide spectral lines which arise in the manner indicated in fig. 77. Sufficiently far out in the atmosphere of a star radiation which is emitted and moves outwards is not absorbed before it escapes from the atmosphere. When a diffuse atmosphere is observed, it is possible to see radiation from matter which is flowing out from both the near hemisphere and the far hemisphere of the star. The former radiation is blueshifted and the latter redshifted relative to the rest wavelength of the radiation and as a result spectral lines, observed from the entire outer layers of a star, are significantly broadened in the presence of stellar mass loss.

Mass loss from massive main sequence stars

Studies in the ultraviolet have also led to the realization that a significant amount of mass is being lost by massive main sequence stars. In the last chapter I have described the evolution of such stars on the assumption that their mass remains constant. Massive main sequence stars have a lifetime of about 10^7 years, which means that if mass loss is to have a significant effect on their evolution it must occur at a rate of order $10^{-6} M_\odot \mathrm{yr}^{-1}$, or even greater in the case of the most massive stars. Such mass loss rates have been observed and this implies that some massive stars can lose as much as half their mass during the main sequence hydrogen burning phase. It is necessary to try to understand what drives this high rate of mass loss given that these stars do not have outer convection zones and that the escape velocity from them is significantly greater than from a red giant of the same or smaller mass.

The process is not fully understood quantitatively but it is believed to involve an interaction of the radiation flowing through the atmosphere of the star with the matter of the atmosphere. To understand this consider two equations from Chapter 3,

$$\frac{dP}{dr} = -\frac{GM\rho}{r^2}, \tag{3.4}$$

$$\frac{dP_{rad}}{dr} = -\frac{\kappa L\rho}{4\pi cr^2}, \tag{3.56}$$

where I am assuming that there is no energy transport by convection. The equations can be divided one by the other to give

$$\frac{dP_{rad}}{dP} = \frac{\kappa L}{4\pi GM}. \tag{7.8}$$

Because $P_{rad} < P$ and both ρ and T decrease outwards in a star, it is clear that the left hand side of (7.8) must be less than unity. If I then apply (7.8) at the surface of the star, I have the inequality

$$L_s < 4\pi cGM_s/\kappa_s, \tag{7.9}$$

where κ_s is the surface value of the opacity.

There is thus an upper limit to the luminosity of a star in which energy transport is by radiation and this limit is approached if radiation pressure is much more important than gas pressure. This is called the *Eddington Limit luminosity*. In massive main sequence stars radiation pressure is becoming important and the luminosity is a significant fraction of the Eddington limit value. Now suppose that a star is losing mass at a rate $\dot{M}_s$. This mass must leave the star with at least escape velocity, so that the energy carried away by the lost mass must exceed

$$L_{\dot{M}} = GM_s\dot{M}_s/r_s, \tag{7.10}$$

where the kinetic energy per unit mass has ben put equal to $v_{esc}^2/2$. If the actual luminosity of the star is a fraction λ of the Eddington limit luminosity, the ratio of (7.10) to L_s is

$$L_{\dot{M}}/L_s = \kappa_s\dot{M}_s/4\pi\lambda cr_s. \tag{7.11}$$

When observed values of M_s, L_s and r_s are used, $L_{\dot{M}}/L_s$ for early type main sequence stars is found to be comfortably less than 1 but nevertheless significant.

The outward flowing radiation possesses enough energy to drive the observed mass loss. The question is, can it do it? Before I discuss this, I can make another comment on the ratio of $L_{\dot{M}}$ to L_s. If I do not use (7.9) this can simply be written

$$L_{\dot{M}}/L_s = GM_s\dot{M}_s/r_sL_s = (\dot{M}_s/M_s)(GM_s^2/r_sL_s). \tag{7.12}$$

In this form the requirement that $L_{\dot{M}}/L_s < 1$ is essentially a statement that the mass loss time-scale must be greater than the thermal time-scale of the star. If mass is to be expelled by calling on the thermal energy supply of the star, there is a limit to the rate of mass loss determined by the rate at which thermal energy is replenished.

Radiation-driven mass loss

I now turn to the question as to how the energy in the outward flowing radiation can be tapped to lead to mass loss. Suppose that in the outer atmosphere of the star there is an abundant atom in the right stage of ionisation and excitation to absorb radiation of a frequency close to that of the maximum emission of the star. When an atom absorbs radiation not only the energy but also the momentum carried by the radiation is given to the atom. When it subsequently re-emits radiation, which it must do, as an atom or ion cannot remain in an excited state for an indefinite time, there is an equal probability of emitting the radiation in all directions. The net effect on all the atoms of the element being considered is that they have gained momentum and are moving outwards. In order for this to lead to mass loss from the star two conditions must be satisfied.

I first note that there has been no net gain of momentum to the element in the entire atmosphere of the star, because matter moving outwards on one side of the star has equal and opposite momentum to that on the other side. If the two sides of the star share their momentum the outward movement will stop. Information travels around the star at about the speed of sound and this will in principle damp down the tendency of mass to move outwards. In practice the virial theorem tells us that the escape velocity from a star is approximately the average speed of sound in the stellar interior, which must exceed the speed of sound at the surface, so that if the escape velocity is approached communication around the star becomes ineffective. The second requirement is that the atoms which are preferentially responsible for absorbing the stellar radiation, and there may be several types of such atom, must share their momentum by collisions with the remainder of the matter in their neighbourhood and that the matter as a whole must still possess a high enough velocity to move outwards. Note that as any absorbing element moves outwards, the wavelength in the stellar radiation, which it is capable of absorbing, is changed by the Doppler effect and this increases the quantity of radiation which it can absorb.

In the atmospheres of massive main sequence stars both hydrogen and helium are effectively fully ionised which means that they are not able to absorb the outward flowing radiation. As a result any radiative driving of mass loss that does occur must be provided by the atoms of the less abundant elements which, as we have seen, contribute about 2% of the mass of the Sun and of stars of similar age. The heavy element content of the globular cluster stars is very much less than that of the Sun. The massive stars in globular clusters (and indeed those of the same age as the Sun) have long since completed their evolution but it seems likely that they did not lose anything like as much mass as the massive stars which can be observed today. It appears that radiative driving of mass loss can be effective in massive stars today although, as I have mentioned, there are still some problems in the fine details of the explanation. Note that it is only stars which emit strongly in the ultraviolet region of the spectrum which provide the appropriate match between the emitted radiation and the absorbing ions. In cooler stars, the visible radiation which is emitted cannot be strongly absorbed because most atoms in their ground states only absorb ultraviolet radiation.

Note that I can refine the discussion following (7.12). Suppose that a magnitude of momentum $\varepsilon L_s/c$ is given to the outward flowing matter per unit time and that the matter just gains escape velocity. Then

$$\varepsilon L_s/c = \dot{M} v_{esc}$$

or

$$M_s/\dot{M} = (1/\varepsilon)(c/v_{esc})(2GM_s^2/L_s r_s). \tag{7.13}$$

Thus the mass loss timescale is at least of order c/v_{esc} times the thermal timescale rather than being of the order of the thermal timescale.

Effects of mass loss on stellar evolution

As I have explained, very high rates of mass loss are observed in massive early type main sequence stars. At first sight it would appear that, as the mass loss occurs right at the beginning of a star's evolution, this must be essentially equivalent to saying that the effective mass of a star is less than its original mass and that star formation is a less efficient process than it originally appeared. This is not, however, quite correct. It is possible to consider two extreme situations. In the first case the mass loss time-scale is longer than the nuclear time-scale of the star. In this case mass loss is unimportant and the star evolves essentially at constant mass. In the second case the mass loss time-scale is significantly less than the nuclear time-scale but longer than the thermal time-scale. In this case the star will evolve down the main sequence as its mass reduces until it attains a mass at which radiative driving of mass loss does not occur. The interesting situation,

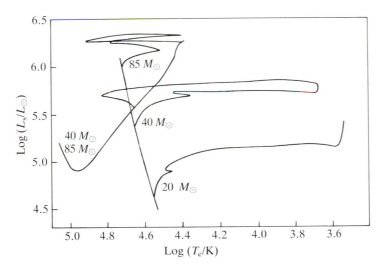

Fig. 78. Evolutionary tracks for massive stars including mass loss. The models have $Z = 0.02$ and the initial masses are indicated on the main sequence. It can be seen that very massive stars may move to the left of the main sequence towards the position of the helium burning main sequence as a consequence of the loss of their outer layers.

however, corresponds to many of the rates of mass loss which are observed. In this case the mass loss time-scale is comparable to the nuclear time-scale, which means that the internal structure and surface properties of the star are neither those of the more massive star nor those of a main sequence star of lower mass. This is illustrated in fig. 78 where I show some theoretical evolutionary tracks of massive stars with mass loss. In some cases most of the hydrogen-rich outer envelope of a star may be lost leaving a helium-burning star close to the helium main sequence. Such stars are believed to be the Wolf–Rayet stars which I mentioned briefly in Chapter 2.

I shall explain in Chapter 10 that there are three possible end points to stellar evolution, *black dwarfs*, *neutron stars* and *black holes*. I shall also explain that there is a maximum mass of a black dwarf which is about $1.2M_\odot$ and a corresponding maximum neutron star mass probably between $2M_\odot$ and $2.5M_\odot$. If no mass loss were to occur all stars more massive than $2.5M_\odot$ would end their lives as black holes. In fact very extensive mass loss does occur and it is, for example, believed that a main sequence star of $5M_\odot$ will become a black dwarf and that one of $25M_\odot$ will become a neutron star. I shall comment on this further in Chapters 9 and 10.

Summary of Chapter 7

Although the existence of explosive mass loss from stars has been recognised for a long time, direct information about steady mass loss from stars in relatively early stages of their evolution has only been available since the development of space astronomy. It is now known that important mass loss occurs both from cool red giants and from massive main sequence stars. Although the mechanisms of mass loss are very different in the two cases, they are both most readily observed in the ultraviolet region of the spectrum. The mass loss is in many cases substantial enough to alter the subsequent evolution of the star. As a result many stars have a final mass which is significantly less than their initial mass.

8

Close binary stars

Introduction

Throughout the book so far I have only discussed the structure of isolated stars, although I have stressed the importance of binary stars in providing information about stellar masses and radii. Although most stars may be partners in binary or multiple systems, for most of them at least for most of their life history the stars are sufficiently far apart that the internal structure of the stars is not affected by the presence of a companion. The mutual gravitational attraction of the two stars causes them to orbit about their common mass centre but otherwise their binary nature can be ignored. This ceases to be the case if binary stars are initially very close or if they become close during their evolution, possibly because one star expands substantially to become a red giant and as a result its surface gets very close to its companion.

When stars are close there are two distinct types of interaction between them. The properties of the surface layers of one star may be affected as a result of irradiation by the other star. This will be particularly true when one component is much more luminous than the other, when the effect on the less luminous star will be very great. In addition both the gravitational attraction of the companion star and the rapid rotation, which must occur because the stars orbit around one another, will cause a star to deviate substantially from spherical symmetry. Ultimately, if the stars are sufficiently close, there may be a transfer of mass and angular momentum and possibly luminosity from one star to the other. The most extreme cases are *contact binary stars* or *common envelope stars*, where two stellar cores share the same atmosphere. When stars are extremely close, and particularly when they are compact stars, *white dwarfs* or *neutron stars*, the evolution may be speeded up by the emission of *gravitational radiation*. The separation of two stars may change during their evolution as a result of mass and angular momentum loss from the system or as a result of transfer between the components. In addition, as mentioned above, one star may expand to become a giant or a supergiant so that its surface approaches its companion. The transfer of mass between components can make the initially more massive star less massive and this can account for the existence of ill-matched binaries in which the less massive star appears to have almost completed its life history, while the more massive star is only just starting. A possible example is the main sequence star Sirius and its white dwarf companion Sirius B, which has only half its mass, although as I shall explain later in the chapter there may be a different explanation in this case. Close binary stars include some of the most energetic events in the Galaxy such as *novae*, some *supernovae* and *X-ray binaries*. For this reason they have an interest and importance out of proportion to their number.

The Roche Model

Most of the mass of a star is concentrated in its central regions. This means that, although a star in a close binary system may have a non-spherical surface, to a close approximation the gravitational field due to the star is still that due to a body with the same mass situated at its centre. The simplest way to discuss a binary system is then to approximate the gravitational field due to each star in this manner. This is known as the *Roche approximation*. The stars must rotate around their centre of mass in such a way that the acceleration of either star in its orbit is balanced by the gravitational field due to the other star. Thus, if a is the distance between the centres of the two stars and a_1 and a_2 are the distances of the two stars from their centre of mass

$$GM_1M_2/a^2 = M_1a_1\omega^2 = M_2a_2\omega^2, \tag{8.1}$$

here M_1 and M_2 are the stellar masses and ω is the orbital angular velocity. Any particle in either star will feel a gravitational attraction due to each of the stars and, in the frame of reference in which the two stars are instantaneously at rest, an additional centrifugal force caused by the rotation. The total force is perpendicular to a surface known as an *equipotential surface*, which has the equation

$$\Phi \equiv \frac{GM_1}{r_1} + \frac{GM_2}{r_2} + \tfrac{1}{2}d^2\omega^2 = \text{constant}, \tag{8.2}$$

where r_1 and r_2 are the distances of the particle from the centres of the two stars and d is its distance from the axis of rotation through the centre of mass of the system. This potential is of course not accurate for particles deep within either star. Equation (8.2) describes what is known as a Roche equipotential surface.

For any value of the constant in (8.2) we have a three-dimensional surface. It is not very easy to visualise such a surface but it is easy to plot two-dimensional sections of it. Consider, for example, what it looks like in the plane which contains the centres of the two stars and the axis of rotation. This is shown schematically in fig. 79. Near either stellar centre the equipotentials are circles corresponding to the principal force being the gravitational force due to one star. The surfaces are distorted from circular shape as the other star is approached and they cross at the point L_1 known as the *inner Lagrangian point*. The part of the surface through L_1

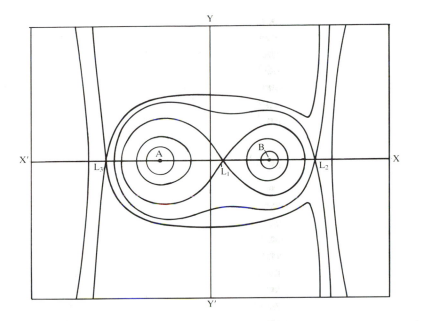

Fig. 79.   Roche equipotentials in the plane containing the centres of the two stars, A and B, and the axis of rotation. The points L_1, L_2, L_3 where equipotentials cross are called Lagrangian points. The equipotential surrounding either star which includes L_1 is called the star's Roche lobe. If a star fills its Roche lobe mass can be transferred to its companion.

which surrounds one star is called its *Roche lobe*. It is this region of the diagram which is most important for our present discussion although there are additional equipotential crossings which are also shown in the diagram.

What is the significance of the equipotential surface which passes through L_1 and which thus encloses both stellar centres? If any material from one star reaches that equipotential, it is energetically possible for it to cross to the other star. If the two stars simply rotate around one another and do not have any spin in addition to that rotation, the surface of each star will be one of these equipotential surfaces. If the two stars are both inside their Roche lobes, the system is called *detached*. If they are well within, the interactions between the two stars are unimportant and, as I have explained, that is the case for most binaries. If the two stars are close to filling their lobes, they will be seriously distorted from spherical shape and there may be important irradiation effects of one star by its companion. If one star fills its Roche lobe but the other is within, we have a *semi-detached system*. In that case mass can be transferred from the lobe-filling star to its companion. As I shall explain shortly, many of the more spectacular stellar events involved semi-detached binary systems. If both stars fill the critical equipotential, we have a *contact binary*, while, if the outer surface of the system is outside the critical equipotential, it is known as a *common envelope system*. All of these types are observed.

A possibly misleading early clue that the evolution of binary stars might be different from the evolution of single stars was afforded by the properties of Sirius and its companion Sirius B. Sirius is a main sequence star of about $2M_\odot$, while Sirius B is a $1M_\odot$ white dwarf. It is not immediately easy to see how they can have reached this state. As I shall explain in Chapter 10, a white dwarf is close to the end point of stellar evolution, whereas Sirius is a main sequence star at a very early stage in its life. Given that more massive stars evolve more rapidly, this is puzzling. An explanation is that Sirius B was originally the more massive star. The original idea was that mass exchange after Sirius B had approached the end of its evolution reversed the mass ratio. This suggestion was made before it was realised that mass loss from single stars is very common, as I have discussed in the previous chapter. Now it seems that straightforward mass loss from Sirius B rather than mass exchange may have made it the less massive star.

Cataclysmic variable stars

One of the best known types of semi-detached binary systems is the *nova*. Related to the classical nova are recurrent novae, dwarf novae and nova-like variables. All of these stars are collectively known as *cataclysmic variable stars*. They are all stars which have outbursts in which their luminosity becomes much greater than its normal value. Dwarf novae have frequent outbursts, recurrent novae have been observed to increase in luminosity more than once, whereas classical novae have only been observed to have one outburst. However, theoretical ideas which I shall now describe suggest that all novae are in fact recurrent but that the recurrence time for a classical nova is very much longer than the period for which they have been observed.

The general picture of a cataclysmic variable is that of a low mass main sequence star filling its Roche lobe with a companion which is a white dwarf. A white dwarf is so much smaller than a main sequence star that the whole star can be assumed to be concentrated at the middle of its Roche lobe. When the main sequence star expands as a result of its evolution or if the stars approach one another because of loss of mass and angular momentum from the system, it overflows its Roche lobe and mass is transferred to its companion. What actually happens is not quite as simple as that because, if the matter leaving the main sequence star conserved its angular momentum, it would not be able to reach the white dwarf. In fact, there will be collisions in the accretion stream which will cause the material to settle down into the plane perpendicular to the rotation axis of the system which contains the centres of the two stars. The matter then forms what is known as an *accretion disk* about the white dwarf. It is then believed that viscous forces in the disk lead to a transfer of angular momentum outwards and mass inwards, so that in a steady state the rate at which mass joins the accretion disk is equal to the rate at which mass falls on to the white dwarf.

This is the picture of a cataclysmic variable between outbursts. The main source of the observed luminosity of the system is the accretion disk with a particular hot spot where the accretion stream joins the disk (fig. 80). The white dwarf is also

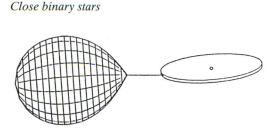

Fig. 80. Mass exchange on to a compact companion. The figure shows the mass-losing star, the accretion stream, the accretion disc with hot spot and the compact star.

observable but in many systems it is very difficult to observe the main sequence star. That is one reason why it was a long time before it was realised that all novae and other cataclysmic variables were binary systems and that this binary nature was absolutely crucial to their properties. The energy radiated by the accretion disk is, in effect, gravitational potential energy. If material could flow directly to the white dwarf, there would be a release of potential energy per unit mass of GM_{WD}/r_{WD}, where M_{WD} and r_{WD} are the mass and radius of the white dwarf. For typical white dwarf masses and radii this is of order 10^{13}J kg^{-1}. Instead of being released at the white dwarf surface, this energy is released and radiated as matter flows through the accretion disk, although the maximum release continues to be close to the white dwarf,

Accretion disks

Accretion disks are believed to be common in the Universe. As was mentioned briefly in Chapter 5, they probably play a role in the formation of stars and planetary systems. In addition it is thought that accretion disks around massive black holes exist in quasars and active galaxies; this has been mentioned in the companion book *Galaxies: Structure and Evolution*. The major problem with present understanding of accretion disks is that the source of the viscosity which enables matter to flow through the disk is something of a mystery. If the gas in the accretion disk is smoothly distributed the source of viscosity is molecular collisions and the magnitude of the kinematic viscosity is approximately $\bar{v}\,l$, where $\bar{v}$ is the average thermal velocity of the gas particles and l is the mean free path of the particles between collisions. Such a viscosity is completely incapable of transporting mass and angular momentum through the disk at the required rate for an understanding of the properties of cataclysmic variables, It is generally believed that the viscosity is caused by matter in quantities much larger than individual particles travelling distances much greater than particle mean free paths before interchanging angular momentum with other elements of matter. One possibility is that the disk is unstable to convection and that the angular momentum is carried by the convective elements. The viscosity, which is much larger that the microscopic particle velocity, goes by the general name of turbulent viscosity.

The simplest model of an equilibrium accretion disk can be described as follows. Because the disk is well within the Roche lobe it is assumed that it is circular around the compact star. This cannot be exactly true because the disk must feel some gravitational attraction from the main sequence star. If the mass of the disk is very much less than that of either star, which is expected to be true, its own self gravitation can be neglected. If in addition the random velocity of the particles of the disk is small compared to the velocity of the rotation around the compact star, the disk is *Keplerian*, which means that the rotational velocity at any point in the disc is given by the equation

$$v_\varphi^2 = GM/r \tag{8.3}$$

in cylindrical polar coordinates (r, φ, z), where M is the mass of the compact star and r is the distance from its centre in the plane of the disk.

If matter joins the disk at a rate $\dot{M}$ and the disk is in equilibrium, the radial velocity of matter flowing through the disk must satisfy

$$\dot{M} = -2\pi r \, \Sigma \, v_r, \tag{8.4}$$

where Σ is the surface mass density of the disk and the radial velocity v_r is negative. The angular momentum per unit mass of matter is $h = rv_\varphi$ and using (8.3)

$$h = (GMr)^{1/2}. \tag{8.5}$$

h is an increasing function of r so that matter can only flow inwards maintaining the equilibrium, if viscosity transfers angular momentum outwards. For the mass transfer rates deduced to be appropriate to disks, the radial velocity is found to be very much smaller than the circular velocity. As a result the energy per unit mass possessed by matter at any radius is essentially the sum of its gravitational potential energy $-GM/r$ and its rotational kinetic energy $GM/2r$ giving a total of $-GM/2r$. As matter flows inward this becomes more negative so that energy is being radiated. It is easy to see that the rate of energy release in a circular shell of the disk between radii r and $r + \delta r$ is $(GM\bar{M}/2r^2)\delta r$. As the area of the shell is $2\pi r \delta r$, the rate at which energy is radiated by unit area of the disk at radius r is $GM\dot{M}/4\pi r^3$ or more accurately half of this amount is radiated from both surfaces of the disk. It can be seen that this increases rapidly as the compact star is approached, so that the major emission from the disk, apart from the hot spot where matter first joins the disk, comes from its inner regions. There is a final emission of energy when the matter joins the white dwarf. The mass of the disk at any time is determined by the value of the viscosity, high viscosity and rapid exchange of angular momentum leading to a low disk mass and vice versa. Once the disk mass is known, the rate at which it must radiate energy determines its temperature and hence the wavelength range in which it primarily radiates. Observations of disk spectra provide estimates of temperatures, masses and viscosity. This is a partial discussion of an equilibrium disk but it is enough for my present purposes.

The outburst

As I have discussed the accretion disk so far, there is no indication that this cannot continue to exist in a steady or slowly varying state as long as mass transfer continues. This is indeed what is believed to be the case in cataclysmic variables between outbursts. What then causes a star to become a nova or one of the related types of star? There are believed to be two different types of outburst one associated with the white dwarf and the other with the accretion disk. I discuss the latter case, which is appropriate to dwarf novae, first.

As I have shown above the energy radiated by a steady accretion disk is directly proportional to the rate of mass exchange between the two stars. A sudden large increase in the disk luminosity must therefore result from a large increase in the mass transfer rate. There are two ways in which this can happen. Either there is an instability in the main sequence star which leads to an increased mass transfer or the accretion disk suffers an instability which causes the mass of the disk to decrease rapidly so that most of its mass falls on to the white dwarf. This could happen as a result of a sudden increase in the disk viscosity. Such an instability would require the disk properties to be slowly varying while the dwarf nova is in quiescence so that the disk can slowly evolve to the unstable state. It is not yet entirely clear which of these mechanisms is the relevant one, although there are some possible tests of them. If the outburst is produced by increased mass loss from the main sequence star, it must start near the outside of the disk and move towards the centre, whereas it is not immediately obvious where in the disk an instability might start. However, there is also a possibility that increased mass loss might be the trigger for the disk instability. If there is increased loss from the main sequence star, this determines the strength of the outburst, while in the disk instability case the outburst strength is limited by the energy release if the entire disk is dumped on the white dwarf.

In the case of classical novae a very different mechanism is believed to occur. As I shall explain in Chapter 10 white dwarf stars must be essentially devoid of hydrogen because their internal temperatures are sufficiently high that, if hydrogen were present, thermonuclear reactions converting hydrogen into helium would produce a luminosity very much higher than observed white dwarf luminosities. As mass is transferred between the two components, the white dwarf acquires a surface chemical composition characteristic of a hydrogen-rich main sequence atmosphere. Eventually this material becomes thick enough that the temperature at its bottom is sufficiently high that thermonuclear reactions can start. It is calculated that thermonuclear runaway can then occur leading to a nova outburst. It is suggested that after an outburst it might take 10^4 years for a critical mass to build up again which is why no classical nova has ever been observed to have two outbursts. It is possible that a classical nova between outbursts might show dwarf nova behaviour.

Magnetic cataclysmic variables

There is a sub-class of cataclysmic variables in which the white dwarf has an intense magnetic field with a surface strength of 10^2T or more. The material in an accretion disk particularly near the compact star is largely ionised and charged particles are constrained to move in helical orbits around magnetic field lines as shown in fig. 81. Let us consider the simplest case in which the magnetic field is approximately dipolar with the magnetic axis coinciding with the rotation axis of the white dwarf and being perpendicular to the plane of the accretion disk. If the field is strong enough, even if the accretion disk forms far from the white dwarf, matter cannot move radially inwards towards the white dwarf. Instead it is channelled along the magnetic field lines to the poles of the star where very intense local accretion and release of energy occurs. This is illustrated in fig. 82.

Fig. 81.   The helical motion of a charged particle about a line of magnetic induction.

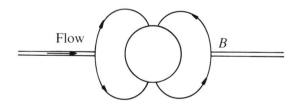

Fig. 82.   Magnetic cataclysmic variable. Close enough to the white dwarf matter is constrained to move along the magnetic field lines and is accreted close to the magnetic poles.

X-ray binaries

In chapter 10 I shall explain that there are two end points of stellar evolution which are even more compact than white dwarfs, *neutron stars* and *black holes*. A typical neutron star has a radius of about 10 km and the gravitational energy released by matter falling to the surface of a neutron star of about a solar mass is in excess of 10^{16}J kg^{-1}. If a main sequence star is the companion of a neutron star and if the main sequence star fills its Roche lobe and an accretion disk forms, the temperature in the disk close to the neutron star is so high that the radiation from the disk is principally in the form of X-rays. When X-ray astronomy was first developed using detectors flown in rockets, binary stars were discovered in which only one component could be seen although there was a strong emission of X-rays from a region close to where the other component was deduced to be. It was suggested that the invisible companion was a neutron star and this was soon accepted as the true explanation. Neutron stars are so small that, unless their surface temperatures are of the order of millions of degrees, their luminosity is too low for them to be directly detected.

When a visible star moves about an invisible star, it is possible to use the methods described in Chapter 2 to estimate the masses of the stars. There will always be some uncertainty in the results because the inclination of the plane of the orbit of the system to the line of sight will not be known. It is however possible to place some constraints on the mass of the compact star. As I shall explain in Chapter 10, there is believed to be a maximum possible mass for a neutron star which is probably between $2.0M_\odot$ and $2.5M_\odot$. If observations of a close binary system indicate the presence of a compact star which is significantly more massive than that, it must be a *black hole*. That certainly seems to be true in the X-ray source known as Cygnus X-1.

Roche lobe overflow is not the only cause of mass exchange between components in close binaries. In some of the X-ray binaries, the main sequence star is a massive star of spectral type O or B. As I explained in the last chapter, massive main sequence stars suffer significant mass loss while they are on the main sequence and some of that mass may be captured by a close companion.

Contact binary stars

Another type of close binary star is the contact binary or *W Ursae Majoris* star named after the first star in the group studied. In these system both stars fill their Roche lobe and in addition there is material in a common equipotential above the surface of the Roche lobes as shown schematically in fig. 83. If I suppose first of all that, apart from rotation of the system about its common mass centre, the system is in hydrostatic equilibrium and that I am only considering parts of the stars for which the Roche potential (8.2) is a good approximation, the equation of hydrostatic equilibrium can be written

$$\frac{dP}{dx} = \rho\frac{d\Phi}{dx}, \qquad \frac{dP}{dy} = \rho\frac{d\Phi}{dy}, \qquad \frac{dP}{dz} = \rho\frac{d\Phi}{dz}, \qquad (8.6)$$

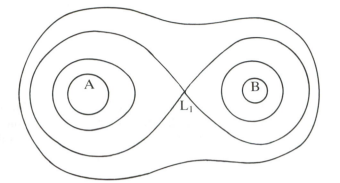

Fig. 83.   In a contact binary star matter fills an exponential surface which surrounds both stars giving them a common outer envelope.

in a cartesian coordinate system. These equations can be written more compactly in the form

$$\text{grad } P = \rho \text{ grad } \Phi, \tag{8.7}$$

which says that the gradient of P is parallel to the gradient of Φ. This can only be possible if the surfaces of constant P correspond with the surfaces of constant Φ which means that I can write

$$P = P(\Phi). \tag{8.8}$$

Equation (8.7) then implies that

$$\rho = dP/d\Phi, \tag{8.9}$$

so that ρ is also a function Φ. Finally, assuming that the mean molecular weight does not vary around an equipotential, the equation of state says that T is a function of Φ. This implies that the surface temperature of the two components must be the same.

It is in fact observed that there is little or any difference between the surface temperatures of the two components of a contact binary system. It is, however, not uncommon for the two stars to have distinctly different masses. If the two stars are main sequence stars or have not evolved far from the main sequence, the luminosity of the more massive star should be significantly higher than that of the less massive star. As normal main sequence stars, they would also have different surface temperatures but, if coming into contact has forced them to have the same surface temperature, the relation

$$L_{s1}/r_{s1}^2 = L_{s2}/r_{s2}^2, \tag{8.10}$$

should be satisfied. The condition that the two stars should fill their Roche lobes gives a relation between r_{s1} and r_{s2} in terms of M_{s1} and M_{s2} and, when this is inserted into (8.10), this provides a relation between L_{s1} and L_{s2}.

The deduced relation between the luminosities is very different from that between the nuclear energy release expected in the two stars. The more massive star is radiating less energy than expected and the less massive star is radiating far more energy than it would be expected to release in nuclear reactions in its interior. The believed solution to this problem is that mass motions redistribute energy between the two components in a common outer convection zone. The existence of these mass motions means that my previous assumption of hydrostatic equilibrium is not quite correct, but this is a correction of detail not of principle. Although it is generally agreed that this is what occurs in the outer layers of contact binaries; it has not proved easy to produce a satisfactory solution to the problem because of the complicated hydrodynamic and thermodynamic processes involved and because of the lack of symmetry in the problem.

Type 1 Supernovae

Some supernova explosions are believed to involve white dwarfs in close binary systems but I will defer discussion of this topic to the following chapter when I describe all types of supernovae.

Summary of Chapter 8

Possibly a majority of stars are members of binary or multiple systems but most of them are sufficiently far from their companions that their evolution proceeds essentially as if they were single stars. There is an important minority of stars which are so close that their evolution is dramatically changed by the presence of a companion. The simplest things that can happen are that a star can be significantly distorted from spherical shape by its companion's gravitational field or that its outer layers can receive substantial irradiation from the companion. If the stars are even closer, mass can be exchanged between them, with obvious effects on their future evolution. The most dramatic effects occur when one star in the system is a very compact star, a white dwarf, a neutron star or even a black hole. In this case mass transfer from a normal star to a compact star involves a large release of gravitational energy, which can be the main source of luminosity in the system in some cases.

The process of mass exchange between two such stars is observed to become non-steady in some cases giving rise to the outbursts of cataclysmic variable stars such as novae and dwarf novae. In most systems the mass is transferred between the components through an accretion disk surrounding the compact star but some white dwarfs possess intense magnetic fields which channel to the mass exchange on to the magnetic poles of the white dwarf.

Another class of close binary is the contact binary or common envelope system in which the two stars share a common atmosphere. Although the two stars may have very different masses, they are observed to have essentially the same surface temperature. This implies that the ratio between their luminosities is not what would be expected from nuclear energy release in their interiors. It is believed that energy transfer occurs between the components in a common convective envelope.

9

Advanced evolutionary phases

Introduction

In Chapter 6 I have given an account of calculations of stellar evolution away from the main sequence. These calculations have not followed a star through its entire life history except approximately in the case of those low mass stars which do not burn their hydrogen and helium before their central temperatures cease to rise and the star as a whole then cools down and eventually ceases to be luminous. For more massive stars there are several evolutionary stages after those which have been discussed in Chapter 6.

It is difficult to calculate the evolution of a star all the way from its initial main sequence state to the end of its life history. There are many reasons why this should be so. One important difficulty is that in all calculations errors tend to accumulate. The numerical processes used in solving differential equations can never be completely accurate and over a long period of integration these mathematical errors tend to pile up. In addition, the mathematical expressions for the physical laws are only approximate. In many cases the physical processes which occur in stars cannot be observed directly in the laboratory and in the case of stellar convection there is neither a good theory nor a good experiment. The uncertainties in the internal structure of a star and particularly in its observed properties may be small when it is on or near to the main sequence, but they might lead to the prediction of an incorrect physical process at a later stage. When I discussed convection in Chapter 3, I indicated that the amount of convective overshooting which occurs in one evolutionary phase might have important consequences later.

Two other factors which could complicate the study of late stellar evolution are rotation and magnetic fields. These are relatively unimportant in most stars on the main sequence but, because of the properties of stellar material, they may become more important at later stages of a star's evolution. The viscosity of stellar material is low and its electrical conductivity is high. This means that as the central regions of a star contract as the star evolves, these regions tend to conserve their angular momentum and also to trap their original magnetic field lines. Conservation of angular momentum and magnetic flux leads to an increase of angular velocity and magnetic induction. In a simple geometrical configuration Br^2 and ωr^2 remain constant where B is the magnetic induction, ω is the angular velocity and r the radius of the region being considered. As contraction proceeds the rotational kinetic energy becomes more important relative to the gravitational potential energy and departures from sphericity will be enhanced.

Stellar instability and mass loss

Perhaps the biggest cause of uncertainty is the amount of mass which is lost by stars during their evolution and precisely when it occurs. I have already explained in Chapter 7 that massive main sequence stars suffer a steady loss of mass and that the rate at which mass is lost is probably related to the heavy element content of the stars. There is not, however, as yet a complete understanding of how much mass such a star will lose before it ends its life, possibly as a Type II supernova or possibly as a black hole. Type II supernovae will be discussed later in this chapter and black holes in Chapter 10.

I have also discussed in Chapter 7 the stellar winds from stars with low surface temperatures and outer convection zones. I discussed the evolution of low mass stars from the main sequence to the helium flash in Chapter 6 on the assumption that the stars evolved with constant mass. It is clear that there is some mass loss from these stars as they ascend the giant branch but once again it is not at present possible to make a precise estimate of the amount of mass lost. Attempts to understand the properties of horizontal branch stars in globular clusters, to be discussed later in this chapter, indicate that they are less massive then their main sequence progenitors but whether all of the mass is lost on the ascent of the giant branch or whether any is lost during the helium flash remains uncertain. As I shall mention later in this chapter, observations of the chemical composition of the surfaces of some red giants also indicates that there has been some process carrying material to the surface from the deep interior.

The mass loss processes which I described in Chapter 7 correspond to a star failing to have a viable equilibrium state. There is also the possibility that an equilibrium configuration is unstable and that the unstable star might lose mass. In order to test for this we ought, at each stage in our calculations, to try to discover what would happen to a star if it suffered a small perturbation such as a compression or expansion in some region or a change of shape. Would the natural tendency be for the compression to increase and for the star to become unstable, or would the compressed region immediately expand to its original state and the star resume its steady evolution? In some cases the physical instabilities may arise without our searching for them. Thus the mathematical inaccuracies which I have mentioned above could introduce the small perturbation of the true solution of the equations which is sufficient to trigger the physical instability. In other cases the equations being used may not allow the instability to arise and it can only be found by deliberate perturbation of the steady equation.

This is true if, for example, normal stellar evolution is proceeding slowly so that (3.4)

$$\frac{dP}{dr} = -\frac{GM\rho}{r^2}$$ (3.4)

can be used. In most cases, if a small imbalance between the two sides of (3.4) is introduced, the star will adjust itself until equality is restored. In other cases the departure from equality will grow and the star will become unstable. Such an instability will only be found automatically if, instead of (3.4), (3.7) is being used:

$$\rho a = \frac{GM\rho}{r^2} + \frac{\partial P}{\partial r}.$$ (3.7)

It might therefore be thought that (3.7) should always be used instead of the approximate (3.4). However, the difference between the two sides of (3.4) is normally so very small that serious mathematical inaccuracies may be introduced by attempting to use (3.7) and it is much safer to test for stability occasionally than to hope to discover instabilities automatically. Of course (3.7) must be used to follow the growth of the instability.

In Chapter 6 I explained that cepheid variables are stars which are unstable to small disturbances. The same is true of RR Lyrae variables and Mira variables in globular clusters. In all of these cases the instability does not grow indefinitely, but instead the star settles down to a state of steady oscillation. This is not the only thing that can happen when a star becomes unstable to a small disturbance. Alternatively the disturbance in the outer layers of a star might grow until the material obtains a sufficiently high velocity to escape from the star. It was at one time believed that there was a maximum possible mass for main sequence stars. Theory predicts that above a certain mass, which depends on stellar chemical composition, the stars suffer a vibrational instability, oscillations of increasing amplitude. The predicted rate of growth of the oscillations is very slow and it is not possible to calculate their build-up through enough oscillation cycles to determine whether they lead to mass loss or to oscillations which are stabilised at large amplitude. Currently it appears that major mass loss may not occur. A somewhat different cause of mass loss is an explosion which occurs in supernovae and (to a lesser extent) in novae. If any mass loss does occur at any stage in a star's evolution, it may alter its whole subsequent life history.

There is another possible effect of instability which is much more difficult to predict theoretically. Suppose there is a very low level instability which goes nowhere near to causing a star to lose mass but which produces internal motions and hence mixing of material inside a star. Because the main sequence phase of stellar evolution is very long-lived, very slow motions could produce significant effects. For the Sun at the present time no substantial changes in its properties occur in less than 10^9 years and a velocity of a few times 10^{-8} ms^{-1} would carry material from the centre to the surface in such a time. It is difficult to be confident that no slow mixing occurs in stars and affects their subsequent evolution.

Neutrino emitting reactions

If stars are being formed at a steady rate in a galaxy, the number of stars observed in any evolutionary phase, except in the final dying phase, can be expected to be proportional to the time which stars spend in that phase. Because the nuclear reactions converting hydrogen into helium release the majority of the energy that can be obtained from nuclear fusion reactions, there are many main sequence stars. The next important nuclear reaction is the conversion of helium into carbon and many giants and supergiants can be identified as helium burning stars. Below a certain mass, stars die before they convert carbon to heavier elements but, as I shall discuss shortly, in sufficiently massive stars nuclear reactions proceed all the way to iron, the most strongly bound nucleus. The energy released by reactions such as carbon burning and oxygen burning is much less than that obtained from hydrogen burning but it is large enough that there should be many stars in these later evolutionary phases. Why is not such a group of stars identified in the HR diagram?

The answer is that when the central regions of a star become sufficiently hot or dense, reactions occur which release neutrinos. The neutrino is such a weakly interacting particle that in most cases it escapes freely from the star adding to the energy loss from the star. Because neutrinos escape from the hot central regions of the star, while photons escape from a much cooler surface, it is possible that the neutrino energy loss can exceed the normal stellar luminosity, in some cases by a very large factor. As a result of this, the time taken in these advanced stages of evolution is very substantially reduced and the probability of being able to identify such stars becomes small. It is likely that they will only be observed in a final supernova explosion, as will be described later in the chapter.

The most important reaction at high temperature is the conversion of an electron and positron into a neutrino and an antineutrino

$$e^- + e^+ \rightarrow \nu_e + \bar{\nu}_e. \tag{9.1}$$

In order for this reaction to occur positrons as well as electrons must be present. In stars with central temperatures of 10^9K or more, electrons and positrons are created from photons by the reaction

$$\gamma + \gamma \rightarrow e^- + e^+. \tag{9.2}$$

The number of electrons and positrons builds up until there are approximately as many of them as photons. At this stage the reverse reaction to (9.2) is operative keeping the numbers in equilibrium. It only requires a very small proportion of the electron/positron annihilations to go to neutrinos instead of photons for this to be the chief energy loss of the star.

At high densities a different process is most important. In an ionised gas, the speed with which photons travel is not precisely equal to the velocity of light. If the electromagnetic field associated with a photon is proportional to $\exp(\mathbf{k} \cdot \mathbf{x} - \omega t)$, the magnitude of $\mathbf{k}$ and ω are related by

$$\omega^2 = k^2 c^2 + \omega_p^2, \tag{9.3}$$

where

$$\omega_p^2 = n_e e^2 / \varepsilon_0 m_e \tag{9.4}$$

and n_e is the number of electrons m^{-3} and ε_0 is the permittivity of free space. ω_p is called the plasma frequency. The energy of a photon is $\hbar\omega$, its momentum $\hbar kc$ and (9.3) takes the form of the special relativistic mass, momentum, energy relation

$$E^2 = p^2 c^2 + m^2 c^4, \tag{9.5}$$

if a mass

$$m = \hbar\omega_p / c^2 \tag{9.6}$$

is attributed to the photon. The theory of the nuclear weak interaction predicts that such an effectively massive photon (denoted Γ) can decay through the reaction

$$\Gamma \to \nu_e + \bar{\nu}_e. \tag{9.7}$$

This plasma neutrino process is very important in dense stars.

When a star is losing a large amount of energy in the form of neutrinos, the equation (3.77) must be replaced by

$$\frac{\mathrm{d}L}{\mathrm{d}M} = \varepsilon - \varepsilon_\nu \tag{9.8}$$

where ε_ν is the rate at which the neutrinos carry away energy per unit mass. If some of the energy release is going to heating up or expanding the element of mass, a similar change must be made to the equation given in the footnote on page 62.

Evolution of low mass stars

I shall illustrate the evolution of low mass stars by considering the stars in globular clusters,because the stars which are currently in an active evolutionary phase in globular clusters are somewhat less massive than the Sun. In fig. 84 I again show the HR diagram of a globular star cluster. As I have described in Chapter 6, direct calculation of the evolution of low mass stars has followed them from the main sequence to the onset of helium burning at the top of the giant branch. The evolutionary tracks and the isochrone shown in Chapter 6 assumed that the stars evolved at constant mass but I have since explained in Chapter 7 that cool stars with deep outer convection zones suffer mass loss. It is therefore now believed that stars in globular clusters are less massive when they reach the helium flash than when they start their ascent up the giant branch. I now need to discuss what happens to stars after the helium flash.

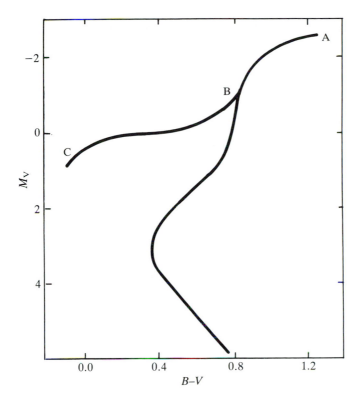

Fig. 84.   The HR diagram of a globular cluster.

In Chapter 6 I discussed how the onset of helium burning in low mass stars is difficult to calculate because the stellar material is degenerate and helium burning starts explosively. The star then evolves so rapidly that it is difficult to follow it accurately. The explosion may lead to the outer regions of the star being blown off into interstellar space. If it is not quite that violent, the material of the star may be stirred up so that the chemical composition of the star becomes more uniform. In any case the internal structure of the star is greatly modified with helium being converted into carbon in a non-degenerate core and with hydrogen burning in a shell further out. When calculations of the helium flash were first attempted, it was assumed that in the initial stages of helium burning the star would move rapidly back down the giant branch to settle on the horizontal branch near B and that it would then evolve towards C. This was, I suppose, the simplest suggestion but it did not have any strong theoretical or observational backing and it now appears that stars are more likely to move very rapidly to the left in the HR diagram to join the horizontal branch near to C and mainly evolve to the right.

If an attempt is made to produce models of horizontal branch stars which lie in the observed position in the HR diagram, it is confirmed that they must be significantly less massive than the stars which are currently leaving the main sequence for the giant branch. Whether or not any mass is lost in the helium flash or whether it is all lost on the ascent up the giant branch is uncertain. Another critical parameter in the comparison of theory with observation is the helium content of globular cluster stars. Unfortunately all of the unevolved stars in globular clusters have such low surface temperatures that no helium spectral lines from which abundances can be detected are observable. All of the stars with higher surface temperatures are ones in which mixing processes might have carried the results of the conversion of hydrogen into helium to the surface. The value of the original helium content of the star affects both its main sequence and horizontal branch properties. Because globular clusters are the oldest known systems in our Galaxy, it is usually assumed that their helium content should be not much different from the primeval helium content predicted by cosmological theory (23–24% by mass) and models of horizontal branch stars are normally calculated with a helium content of about 25%.

The morphology of the horizontal branch varies greatly from cluster to cluster. In particular there are substantial differences in the relative numbers of stars at the red and blue ends of the horizontal branch and in the number of RR Lyrae variables, which populate a region of the horizontal branch between C and B. Calculations similar to those described for cepheid variables in Chapter 6 suggest that stars in the region of the HR diagram populated by the RR Lyrae stars should be variable, but this does not explain why some globular clusters contain more than a hundred of these variables while other clusters with a similar total number of stars contain virtually no variables. There are significant differences in the heavy element content of globular clusters, even though in all cases the content is much less than that of the Sun. It is possible that heavy element content might have a significant effect on the occurrence of variability or on the time spent in the region where variables occur. It is certainly true that M92, a globular cluster with a very low heavy element content, has very few variables, while M3, which has more heavy elements, has more than a hundred variables.

The asymptotic giant branch – thermal pulses

When the stars have moved from C to B along the horizontal branch, calculations predict that they climb up the giant branch again, provided that they are massive enough; they are then said to be on the *asymptotic giant branch*. At the end of their second ascent of the giant branch, which is probably higher than their first ascent, they might be expected to start burning carbon in their central regions. In the first edition of this book, I expressed the view that the onset of carbon burning in the degenerate cores of globular cluster stars would lead to an explosion and mass loss leading to formation of planetary nebulae. I also suggested the explosive carbon burning in some single stars might produce Type I supernovae. Now neither of these is accepted as the true explanation. I shall now explain why that is the case for globular cluster asymptotic giant branch stars and I shall discuss supernovae later in the chapter.

Asymptotic branch stars are believed to produce planetary nebulae but the cause is believed to be *thermal pulses*, which occur before the onset of carbon burning and, indeed, prevent the carbon from being ignited. I have previously explained that, whereas the onset of nuclear reactions in a degenerate gas is likely to be explosive, there are unlikely to be any problems in a non-degenerate gas. Although this is generally the case, particularly in stellar cores, it has been discovered that instability can occur in a thin nuclear shell source, even when the gas is non-degenerate. This arises in the case of helium shells because of the very high temperature dependence of helium burning which is approximated in (4.28). Calculations have shown that such instabilities, which have been given the name of thermal pulses occur in asymptotic giant branch stars and that they can terminate with a phase of substantial mass loss, which produces a planetary nebula and which leaves a remnant star which is incapable of burning carbon. The mass ejected is ionised by the hot central star and is visible as a planetary nebula until expansion and cooling leads it to become part of the general interstellar medium. The central stars are believed to contract and cool and to become white dwarfs.

Planetary nebulae and white dwarfs

The possible connection between planetary nebulae and white dwarfs is shown in fig. 85. An evolutionary track has been calculated for a star of $0.6M_\odot$, whose only sources of energy are gravitational energy release and cooling. It can be seen that the evolutionary track passes through the regions in the HR diagram occupied by both the central stars of planetary nebulae and white dwarfs. Although $0.6M_\odot$ is rather lower than the mass usually estimated for stars in active

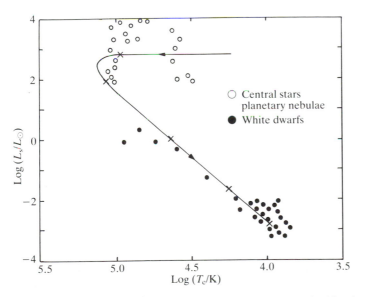

Fig. 85.   The HR diagram for nuclei of planetary nebulae and white dwarfs; also shown is the evolutionary track of a star of $0.6M_\odot$.

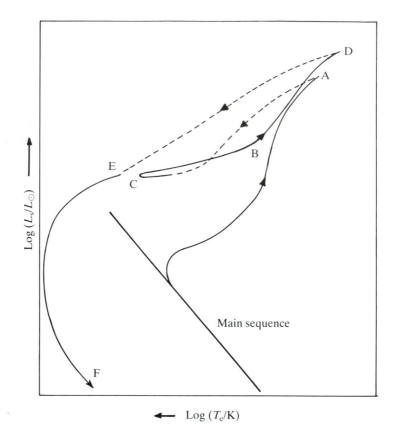

Fig. 86.   Schematic evolutionary track of a star of low mass.

evolutionary phases in globular clusters, it is probably not inconsistent with these, if the mass loss accompanying the formation of a planetary nebula is taken into account. Some of the neutrino reactions mentioned earlier in the chapter accelerate the early cooling of a central star towards a white dwarf.

Figure 86 shows a possible complete evolutionary track (post-main-sequence) for a star which is massive enough to burn helium and to have thermal pulses, but which never becomes hot enough in its centre for nuclear reactions leading to the most strongly bound nuclei in the neighbourhood of iron in the periodic table. It should be stressed that this diagram is only schematic; in particular it does not include small excursions due to thermal pulses. Different regions of the diagram have been explored, but the evolution of a single star through all of the phases shown has not been studied. The dashed regions of the curve between A and C and D and E represent rapid dynamic phases of evolution which have hardly been explored by direct calculation and they are only meant to indicate that the stars must get from A to C and from D to E somehow. One planetary nebula has been found in a globular cluster (M15) and its central star appears to lie a little above the horizontal branch in its cluster HR diagram, which is consistent with the above schematic evolutionary track.

Surface chemical composition of giants

One of the most important predicted and observed properties of giant stars is that it is possible for products of nuclear reactions inside the stars to be brought to the surface. The first dramatic success of ideas concerning stellar nucleosynthesis was the discovery in 1952 of spectral lines of the element technetium in an S type red giant. Technetium has no stable isotopes and its presence in the stellar atmosphere can only be explained if it was produced in the star. I have explained in Chapter 4 that, when the CN cycle of H burning is operating in equilibrium, there is conversion of carbon to nitrogen. In addition there is an increase of ^{13}C relative to ^{12}C. The observation of relative abundances comparable to the CN cycle values, rather than the very different ratios observed in the Sun and most stars, is evidence that the CN cycle has operated in the star being observed.

Red giant stars have deep outer convection zones and, in order to obtain anomalous surface abundances, it is necessary for the convection zone to reach a region in which nuclear reactions have occurred or to connect with a region in which products of nuclear burning have previously been mixed. A good example is what happens in a thermal pulse cycle. The asymptotic giant branch stars sometimes have hydrogen burning in a shell source. When the helium, deeper in the star becomes hot enough for a helium shell flash, the hydrogen burning ceases and, at the same time, the helium burning shell has a convection zone above it in which the products of helium burning are mixed. Later, when the helium burning has become more quiescent, the star has a deep outer convection zone which is able to *dredge up* products of helium burning to the surface.

Supernovae

The most dramatic endpoint of stellar evolution is the explosion of a supernova. When this happens, for a period of weeks or months a single exploding star may have a luminosity comparable with that of a small galaxy. There are two principal types of supernova, Type I and Type II, although these classes, particularly Type I, contain subdivisions. For our present purposes, the main distinction between Type I and Type II supernovae is that the former are observed in all classes of galaxy, and in all regions in galaxies, whereas Type II are only found in regions of active present star formation. It is an obvious inference that Type II supernovae must be massive stars, which can evolve to the end-point of their evolution in not more than a few times 10^7 years. Conversely, it appears that Type I supernovae must be low mass stars because only low mass stars are currently luminous in some of the galaxies and regions of galaxies in which they are observed. This may not be true of all the sub-classes of Type I supernovae, but in any case at least two causes of a supernova explosion must be found.

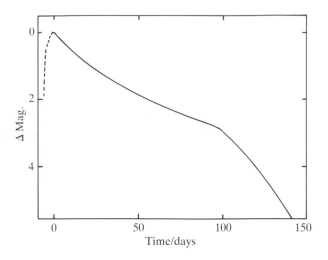

Fig. 87.   Light curve of a supernova of Type II. The rise to maximum light shown by the dashed section is not usually observed.

Supernovae cannot radiate at their high rate for very long and they are very bright for only a few months and then they become faint and eventually invisible. The light curve of a supernova is shown in fig. 87. At present not a great deal is known observationally about pre- and post-supernovae. The last supernova seen in our Galaxy was in 1604, which just preceded the use of the telescope in astronomy. There was also one in 1572 and these two are known as Tycho's star and Kepler's star because of the careful study of them made by Tycho and Kepler. Three more supernovae in our Galaxy in the past 1000 years have been identified from Chinese records. Many more supernovae have been observed in galaxies other than our own since the late 1930s and from these observations it has been estimated that, on the average, there might be one supernova per (large) galaxy in about 30 years. There is probably not a serious discrepancy with the smaller frequency of observed supernovae in our Galaxy because, for much of the disk of our Galaxy, interstellar absorption is so strong that not even a supernova would be readily seen. A few neutron stars, detected as pulsars, have been found in supernova remnants. Our knowledge of supernovae has been greatly increased by the explosion of supernova 1987A in the Large Magellanic Cloud, a satellite galaxy of our own, and I shall discuss that briefly later in the chapter.

Supernova of Type Ia

These are the main sub-class of Type I supernovae. As I stated earlier it used to be thought that the onset of carbon burning in a single degenerate star was the cause of a Type I supernova explosion, but it is now believed that many, if not all, Type I supernovae occur in binary systems. In the case of Type Ia the explosion is believed to follow mass transfer from a binary companion on to a white dwarf. Whereas in the novae discussed in Chapter 8, such a mass transfer leads to hydrogen burning and a small mass loss, the white dwarfs in this case are believed to be composed of carbon and oxygen. One popular scenario has a binary composed of two CO white dwarfs which merge and, as a result of the merger, explosive burning of carbon and oxygen occurs leading eventually to the production of ^{56}Ni and other iron peak nuclei. It is thought that the early light curve of these supernovae, which are all very similar, is dominated by the radioactive decay of ^{56}Ni to ^{56}Co and ^{56}Fe. I shall discuss this process further in connection with SN1987A.

The late evolution of massive stars and Type II supernovae

I have mentioned several times that when a star becomes degenerate it is possible that its central temperature may reach a maximum value and that the star may subsequently cool down and die (see for example page 108), and this is what happens in the case of the globular cluster stars which I have described above. The lower the mass of a star the earlier in its evolution this is likely to happen. For sufficiently massive stars, the central regions may not become degenerate until their evolution is essentially over. I can imagine such a star passing through a succession of evolutionary stages in which first one nuclear fuel and then another nuclear fuel supplies the energy liberated in the central regions of the star and contributes to the energy which is radiated from the star's surface. As each successive fuel supply is exhausted in the centre of the star, these central regions contract and heat up until the next series of nuclear reactions becomes operative. I have not discussed in Chapter 4 the full range of energy-releasing nuclear reactions which can occur in stars, but I have said that no further energy can be obtained from nuclear fusion reactions once the material has been converted into nuclei in the neighbourhood of iron. As I have explained earlier in his chapter, the final nuclear reactions in such a star will not supply the energy which is radiates for very long by astronomical standards, because of the copious production of neutrinos in the central regions of the star.

Of course this conversion to iron will not happen simultaneously throughout the whole star but initially the centre of the star will reach that state. The variation of chemical composition in a highly evolved massive star might appear schematically as shown in fig. 88. When the central regions are mainly composed of iron, they must contract again with no hope that further nuclear reactions will provide the energy which will flow down the temperature gradient, which must still exist, if the central regions of the star are to have a pressure gradient capable of balancing the inward gravitational force. Eventually the collapse will bring the central regions to a sufficiently high density that the electrons become degenerate and it might be thought that the contraction would then stop and that the star would cool down like the low mass stars I have described earlier. In fact, in the next chapter, I shall show that there is a maximum possible mass for a cold degenerate star, which is much less than that of stars which reach the end-point of nuclear evolution. This remarkable result means that even the pressure of degenerate electrons (or

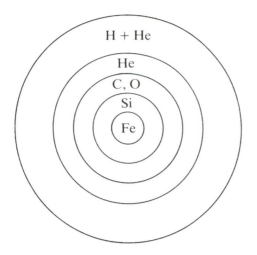

Fig. 88.   Schematic chemical composition of a highly evolved massive star.

neutrons) cannot stop a massive star contracting and enable it to cool down and die quietly. In turn this suggests that if nothing else intervened, such a star would collapse to a black hole. I shall explain that it is believed that a supernova explosion may intervene.

When the central regions of a star are composed of elements such as chromium, manganese, iron, cobalt and nickel the so-called iron group of elements, does this mean that no further nuclear reactions will occur in the centre of the star? There are no longer any nuclear reactions which release nuclear binding energy in significant amounts. However, if the particles in the stellar material have high kinetic energy, which they do when the temperature is high, there is no reason why nuclear reactions should not occur in which the kinetic energy of the particles is used to enable less strongly bound nuclei to be produced. This is what happens in many artificially induced nuclear reactions in the laboratory when particles, which have been accelerated to very high energies, hit a stationary target. As the temperature at the centre of the star increases, a situation is reached which is analogous to what happens when an atomic gas is heated. Energy can be released by forming atoms out of ions and electrons and energy is required to separate electrons and ions, but despite this the gas becomes ionised when its temperature is raised. This ionisation (and also molecular dissociation) is what happens in the early stages of the evolution of a protostar described in Chapter 5. In a similar way, when the material in the centre of a highly evolved star exceeds about 5×10^9K, the nuclei will be dissociated, eventually being converted into a mixture of protons and neutrons.

The dissociation of the most abundant isotope of iron would take the form

$$^{56}\text{Fe} \rightarrow 13^4\text{He} + 4\text{n} \tag{9.9}$$

and this would be followed at a somewhat higher temperature by

$$^4\text{He} \rightarrow 2p + 2n. \tag{9.10}$$

Thus the whole range of nuclear evolution, which has been so painstakingly followed in the earlier states of the star's evolution, is essentially reversed. This reversal requires a large supply of energy which initially comes from the kinetic energy of the particles. Left to itself this would cause the temperature and pressure to fall, but the reduction in pressure means the matter is less able to resist the gravitational force and the central regions collapse very rapidly and continue to heat up. A similar rapid collapse occurs in protostars at the time of hydrogen dissociation and ionisation, as I have discussed in Chapter 5. This collapse of the central regions means that, effectively, it is gravitational potential energy which converts the iron peak elements back to protons and neutrons. In the process the central region of the star may be so compressed that its constituent protons and electrons are forced to combine to form even more neutrons through the reaction

$$e^- + p \rightarrow n + \nu_e. \tag{9.11}$$

This provides an additional flux of neutrinos, but not antineutrinos, from the star in addition to the neutrino/antineutrino pairs produced by reaction (9.1) and similar reactions producing muon (ν_μ, $\bar{\nu}_\mu$) and tauon (ν_τ, $\bar{\nu}_\tau$) neutrino pairs. As a result the star may develop a core of closely packed neutrons. This will be discussed further shortly and in Chapter 10.

Theoretical models of Type II supernovae

I now have to explain why in at least some cases the collapse of a massive star is followed by a supernova explosion leaving behind a neutron star as a remnant. This is believed to happen for the lowest mass stars which have the evolution described in the previous paragraph. Even more massive stars are thought to leave a remnant which is a black hole and the possibility exists that some stars might become a black hole without a supernova explosion. Because the extent of mass loss at earlier stages of stellar evolution is not fully understood, there is not yet a clear relation between the mass of a star at the time it becomes a supernova and its main sequence mass. In any case, as mentioned in Chapter 7, this is bound to depend on the heavy element content of the star.

The attempt to understand Type II supernova explosions has taken a long time. When the first serious studies were made in the 1960s, it was assumed that the cause of the explosion was thermonuclear. The collapse of the core of a star leads to material further out in the star, which contains potential nuclear fuel, falling inwards and being heated rapidly. As a result rapid nuclear reactions can occur and blow off the outer layers of the star. Although the nuclear fuel appeared to contain enough energy for this to happen, detailed calculations of such a thermonuclear explosion never demonstrated that the star would be disrupted. Rapid thermonuclear reactions *must* occur at the time of a supernova explosion but it appears that they do not cause it.

The next suggestion came from the realisation that, in the final stages of the collapse of a stellar core, it could become opaque to neutrinos because of the great increase in the mass per unit area that the neutrinos have to pass through to escape. It was suggested that the energy or momentum possessed by the neutrinos could be used to blow off the outer layers of the star above the level at which they are absorbed. In a sense this would be analogous to the radiation pressure driven mass loss discussed in Chapter 7. Once again detailed calculations did not provide confirmation of an explosion. Instead it became accepted that inside the core of a collapsing star there would be a region from which neutrinos could not escape freely and that they would in effect be radiated from a surface called the neutrinosphere, which is analogous to the photosphere or visible surface of a star.

The currently favoured mechanism is rather more mundane than the previous two. Calculations of the pre-supernova evolution indicated that the mass of the central core tended to be relatively independent of the total mass of the star, being about $1.4M_\odot$ for quite a large range of total mass. As this mass attempts to become a neutron star, of the type to be described in the next chapter, it becomes effectively incompressible. The sudden change from a rapidly contracting core to one that cannot contract any further leads to a bounce of material falling on to the compact core and it is this bounce which is supposed to send a pressure wave outwards which removes the outside of the star.

Supernova 1987A

In February 1987 astronomers observed the explosion of a supernova in the Large Magellanic Cloud. At a distance of only about 50kpc from the Earth, this is by far the nearest supernova to have been seen since the invention of the telescope. In addition theoretical ideas about supernovae and observational techniques have both developed to an extent that the observations that could be made could be fully exploited. Because of the nearness of the supernova, it became possible for the first time to identify the star which became the supernova by studying photographs of the same region of the sky taken before and after the explosion. The supernova is classified as one of Type II although it is atypical of that type. The precursor star was a blue supergiant although previous ideas had expected pre-supernovae to be red supergiants. This discrepancy probably arises from the mass loss history of the star. Many things have been learnt from a study of supernova 1987A but I will mention just two of them.

There has for some time been a need to account for what happened to the gravitational energy released in the formation of a neutron star in a stellar collapse. If the star settles down as a static neutron star, the energy must be radiated somehow. The total electromagnetic energy radiated by a supernova and the total kinetic energy of the mass expelled in a supernova explosion can be estimated and compared with this gravitational energy release but typically they only account for about 1% of it. What has happened to the remainder of the energy? I have already referred to the great energy loss in the form of neutrinos and antineutrinos during stellar collapse and it appears that this could account for the remainder of the energy. Can this be tested?

By chance, at the time of the explosion of Supernova 1987A, there were some experiments in operation trying to test the prediction of some particle physics theories that the proton might be unstable with a half-life significantly in excess of 10^{30} years. The apparatus used was also capable of detecting electron antineutrinos and a pulse of these was measured a few hours before the appearance of the visible supernova. This is what would be expected. The neutrinos are radiated at the final stages of stellar collapse and travel towards the observer at the speed of light whereas the rise to maximum light of the supernova only occurs as the outer layers of the star expand and the area of the photosphere is greatly increased.

Although only about twenty neutrinos were detected, a considerable amount of useful information was obtained. Knowing the size of the detectors and the distance of Earth from the supernova it is possible to estimate whether the number of neutrinos detected and the energy which they individually possessed is compatible with the idea that most of the gravitational energy released is converted into neutrinos. When allowance is made for the electron neutrinos and the muon and tauon neutrinos and antineutrinos which could not be detected, it appears that observation and theory are consistent within their errors. In addition this observation places an upper limit to the mass of the electron neutrino. If a neutrino has a very small mass its velocity is not quite equal to c and the velocity depends on the energy of the neutrino. The spread of neutrino arrival times at Earth can arise from three sources; spread of times of emission, differing travel times from different points in the pre-supernova and variation in neutrino energy. If the whole spread of travel times is attributed to neutrino mass, which cannot be correct, an upper limit to the neutrino mass of order $15 eV/c^2$ is obtained. This is a remarkable example of how astronomy can be used to provide results in fundamental physics.

The second comment refers to the main source of the early luminosity of the supernova. Many years ago it was noted that the early decay phase of a supernova's luminosity resembled that of a radioactive decay and it was suggested that decay of some isotope produced in the explosion powered the luminosity. The original suggestion was ^{254}Cf which has been observed in atomic bomb tests but it was really implausible that sufficient ^{254}Cf would be produced to dominate a supernova's emission. It was subsequently realised that the build up to iron group nuclei would most plausibly occur through α particle nuclei, nuclei with equal even numbers of protons and neutrons. In that case the most abundant nucleus produced initially would be the unstable nucleus ^{56}Ni which would subsequently decay to ^{56}Fe through the reactions

$$\left. \begin{array}{l} ^{56}\text{Ni} \rightarrow {}^{56}\text{Co} + e^+ + \nu_e \\ ^{56}\text{Co} \rightarrow {}^{56}\text{Fe} + e^+ + \nu_e \end{array} \right\}, \tag{9.12}$$

with half lives of 6 days and 77 days respectively. It seemed possible that nuclear reactions close to the time of the explosion would produce sufficient ^{56}Ni to provide the bulk of the luminosity.

Theoretical calculations suggest that SN1987A was a star of about $20M_\odot$, which underwent a Type II explosion. It is not entirely clear why the star was a blue supergiant rather than a red supergiant and there continue to be some difficulties in understanding the explosion mechanism. This is not really surprising when it is realised that only an odd per cent of the available energy goes into the outburst. The production of cobalt to the extent of about $0.07M_\odot$ has been observed but it is not distributed symmetrically in the remnant. It appears that some instability has caused spikes of Co to penetrate into the region occupied by elements lower in the periodic table. It is clear that, even if nuclear reactions do not cause a supernova explosion, they do influence what is seen. Supernova 1987A has answered some of the questions relating to supernova explosions but it has asked many more.

Origin of chemical elements and cosmic rays

There are several reasons why there is great interest in supernovae other than the ones mentioned above. It is generally accepted that the original composition of the Galaxy was very simple. The *hot big bang cosmological theory*, which is in good general agreement with observations of large scale structure of the Universe, suggests that the pregalactic composition should have been essentially a mixture of hydrogen and helium. In turn it is suggested that the heavier elements have been produced by nuclear reactions in stars. If this is so, the heavy elements which are observed in stars today must have been produced in previous generations of stars and have been subsequently expelled into interstellar space. That explains why there is great interest in mass loss from stars, apart from its effect on stellar evolution. When mass loss occurs it is very important. If it occurs when a star is on or near to the main sequence, the matter which is expelled will have the same composition that it had when the star was formed and this will make no contribution to the origin of the elements. Mass loss at late stages of stellar evolution is more effective, but in the low mass stars, such as those that form planetary nebulae, nucleosynthesis does not go beyond the products of helium burning. It is in a supernova explosion that there is a large loss of mass from a highly evolved star. It appears that most of the heavy elements in our Galaxy must have been produced in stars which became supernovae.

The *cosmic rays* may also be associated with supernovae. Cosmic rays reach the Earth almost equally from all directions. They are very high energy particles travelling with velocities almost equal to that of light and some origin of these high velocities has to be found. Individual cosmic rays have been found with energies up to 10^{20}eV. It appears that the highest energy cosmic rays may be produced outside our Galaxy, possibly in massive outbursts such as are observed in some galactic nuclei, but it seems that most of the cosmic rays must be produced in the Galaxy. It is then natural to associate their production with the most violent events known to occur in the Galaxy. There is direct evidence of the existence of high energy particles in supernova remnants such as the Crab Nebula which I discuss in the next paragraph. Radio emission from the remnants is identified as *synchrotron emission*, which is radiation from relativistic electrons gyrating in a magnetic field. There are theories which can account for the acceleration of particles to relativistic velocities in the remnants themselves. In addition the neutron stars left behind by some supernovae are observed as *pulsars*, which I shall discuss in the next section and in Chapter 10. The immediate surroundings of a pulsar also appears to be a suitable environment for the acceleration of particles to high energies using the electromagnetic fields associated with the pulsars.

The Crab Nebula

One of the most remarkable objects known to astronomers is the Crab Nebula. There are probably many other objects equally remarkable but none of them is sufficiently close to us to be studied in the same detail. The Crab Nebula is in the site of a supernova seen by the Chinese in the year AD 1054† and the nebula is believed to have been produced by the supernova explosion. Optically it is a complex of bright gaseous filaments separated by dark regions. The filaments are expanding in a way which is consistent with their having been thrown off in an explosion some 900 years ago. The Crab Nebula provided a major puzzle for astronomers. The total emission of the nebula throughout all of the electromagnetic spectrum, from the radio to X-rays, amounted to 10^{31}W or about $2.5 \times 10^4 L_{\odot}$. How could a supernova be putting this amount of energy into its remnant 900 years after it exploded? In addition, although a number of stars could be observed in the direction of the nebula, it was not clear that any one of them was in the nebula and was the remnant of the supernova.

All of this was changed by the discovery of the Crab Nebula pulsar in 1968. This is a source of radio waves which emits pulses of radiation every 0.03s. The pulsar was subsequently shown to coincide in position with one of the stars in the direction of the nebula and it was then found that all of the optical luminosity was emitted in pulses with the same period as the radio pulses. The star was obviously a dead or dying star with no measurable normal luminosity. Pulses are also found in X-rays and γ-rays. As observations of the pulsar accumulated, it was found that the period of the pulsar is lengthening slowly with a characteristic time of 1000 years for significant change. This ties in with the idea that the pulsar is a remnant of the supernova explosion of 900 years ago. As will be explained in the next chapter, it is believed that pulsars are rotating neutron stars. The kinetic energy lost by the slowing down of the Crab Nebula pulsar is available to power the nebula and a figure of 10^{31}W seems perfectly reasonable. This provides strong evidence that a supernova explosion can leave behind a neutron star and that the consequences of the explosion can be observed many years later. Astronomers would be very excited if there were to be another supernova in our Galaxy and one nearer than the Crab; though not too near!

Summary of Chapter 9

Some stages in stellar evolution have not yet been reached by direct evolutionary calculations from the main sequence. Such direct calculations may be difficult for several

---

† The Chinese records of novae and supernovae have proved very valuable. They called them guest stars. The value of careful observations to posterity can never have been demonstrated more clearly than in this case.

reasons. Errors tend to accumulate and this may make results very uncertain. In some cases stars become grossly unstable and the resulting periods of mass loss are difficult to study reliably. Even very low level instabilities might lead to mixing of material inside a star with effects on later evolution. On the observational side stars in late evolutionary stages may be difficult to observe because the stages are so short-lived. Stellar lives are shortened by neutrino-emitting reactions at high temperatures and densities.

In the case of low mass stars, the first natural break, apart from some uncertainty of mass loss on the giant branch, occurs at the onset of helium burning. After the helium flash, which may involve some further mass loss, the star settles down on the horizontal branch in the HR diagram, but the details of the transition are still not clear. Further evolution takes the star into the giant region again and a second break in direct calculations occurs when thermal pulses initiate mass loss which leads to the formation of a planetary nebula and which leaves a remnant star which cools to become a white dwarf. At several stages in the evolution of low mass stars, matter which has been involved in internal nuclear reactions is carried to the surface and the appearance of spectral lines of some elements and isotopes in observed spectra confirms that this has happened.

The most dramatic event in stellar evolution is the explosion of a supernova. Many Type I supernovae are believed to involve white dwarfs in close binary systems, with the explosion powered by rapid burning of carbon and oxygen. Type II supernovae are, in contrast, believed to arise from massive stars whose central regions have been converted into iron and neighbouring elements. No further energy can be released by nuclear reactions in these central regions, but they continue to increase in temperature. Eventually the iron is converted back to helium and neutrons, and the energy required for this conversion is provided by rapid collapse releasing gravitational energy. Finally the core is converted to a neutron star which cannot collapse any further and a supernova is produced when the outside of the star bounces off the core and is expelled into space. Most of the gravitational energy which is released comes out in neutrinos rather than photons and mass motions. Much of the early luminosity of a supernova may be provided by radioactive decay of ^{56}Ni produced just before the outburst. The neutron star remnant may be a pulsar and the origin of the heavy elements and of cosmic rays is associated with a supernova explosion.

There are still considerable uncertainties in the late stages of stellar evolution but I hope that what I have described is true in broad outline if not in fine detail.

10

The final stages of stellar evolution: white dwarfs, brown dwarfs, neutron stars and black holes

Introduction

In the previous discussion of stellar evolution it has frequently been remarked that, so long as the stellar material remains in the form of an ideal classical gas, its central temperature can only increase as it evolves. This result was originally deduced from the Virial Theorem (3.24) on page 55. As I have mentioned on page 201 in Chapter 9, there is at present no completely clear solution to the problem of what happens to a star whose central temperature is still rising at the time that nuclear fusion reactions have converted the central regions to iron, although the association with supernovae of type II seems highly probable; in fact, as I shall explain in the last section of this chapter, the problem can arise even earlier than that. However, if the centre of the star ceases to be an ideal classical gas and becomes a degenerate gas, it is possible that the central temperature may pass through a maximum and that the star may cool down and *die*. This possibility has already been illustrated for low mass stars in figs. 76 and 86 Such a dying star is likely to have a low luminosity. It is also likely to have a high density. It can only begin to cool down after its central regions have become degenerate and, if the central temperature has previously risen sufficiently for one or more sets of energy-releasing nuclear reactions to occur, a very high density is necessary before degeneracy can occur, as has been seen in Chapter 4 (fig. 45)

Such under-luminous dense stars have been oberved. They are the white dwarfs which have been discussed briefly in Chapters 2 and 8. All white dwarfs are of low luminosity and those which are partners in binary systems, for which masses are known, are of very high density. Those whose masses cannot be determined almost certainly have a very small radius, as will be discussed on page 212 and it follows that they also have high densities unless their masses very small indeed. The white dwarfs which have been observed do not have particularly low surface temperatures whereas we might expect dying stars to exist with very low surface temperatures. However, as is shown in fig. 89, the luminosity of known

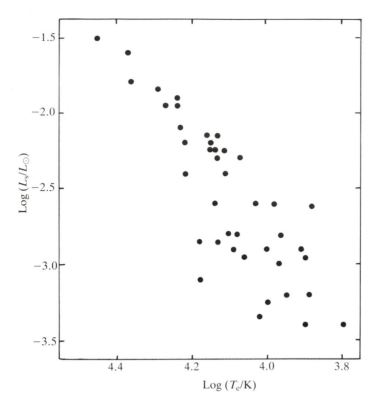

Fig. 89.   HR diagram for white dwarfs.

white dwarfs decreases with surface temperature. This means that, even if white dwarfs exist with much lower surface temperatures, they are likely to be too faint to be observed, even if they are quite near to the Sun. In addition there is a limit to the extent to which a white dwarf can have cooled, even if it was formed very early in the life of the Galaxy. It is believed that this is why no very faint and cool white dwarfs have been observed and the lower limit of white dwarf luminosity has been used to obtain an estimate of the age of the disk of the Galaxy.

The structure of white dwarfs

Dying stars must be degenerate in their central regions and from fig. 45 it can be deduced that, as they cool down, they must become increasingly degenerate. Because of this, as a first attempt to study the structure of white dwarfs, I investigate the properties of stars which are made of degenerate gas throughout. In an actual star there is likely to be at least a thin surface layer in which the material is nearer to an ideal classical gas. In Chapter 4 I have given two formulae for the pressure of a degenerate gas. These are the non-relativistic formula:

$$P_{\mathrm{gas}} \simeq K_1 \rho^{5/3}, \tag{4.49}$$

which is valid when the maximum momentum p_0 of the electrons satisfies $p_0 \ll m_e c$ and the relativistic formula:

$$P_{gas} \simeq K_2 \rho^{4/3}, \tag{4.51}$$

which holds when $p_0 \gg m_e c$. I have also stated in Chapter 4 that there must be a gradual change from the first of these formulae to the second as p_0 and ρ increase and the general form of the pressure-density relation can be shown to be:

$$P_{gas} = f\{(1 + X)\rho\}, \tag{10.1}$$

where f is a function which changes from $K_1 \rho^{5/3}$ at low values of ρ to $K_2 \rho^{4/3}$ at high values of ρ and X is, as usual, the fractional hydrogen content by mass.

The structure of such a fully degenerate star can now be studied by solving the equation of hydrostatic support:

$$\frac{dP}{dr} = - \frac{GM\rho}{r^2}, \tag{3.4}$$

and the equation of mass conservation:

$$\frac{dM}{dr} = 4\pi r^2 \rho, \tag{3.5}$$

in conjunction with (10.1). Because the pressure given by (10.1) is independent of the temperature, these equations form a complete set which can be solved without considering the thermal structure of the star. By this, I mean that it is not necessary to study how the temperature varies in the star or how energy is transported. The differential equations (3.4), (3.5) are best considered once again in the form:

$$\frac{dP}{dM} = - \frac{GM}{4\pi r^4}, \tag{3.75}$$

$$\frac{dr}{dM} = \frac{1}{4\pi r^2 \rho}, \tag{3.76}$$

when the boundary conditions to be applied are:

$$\left. \begin{array}{ll} r = 0 & \text{at} \quad M = 0, \\ \rho = 0 & \text{at} \quad M = M_s. \end{array} \right\} \tag{10.2}$$

If values of X and M_s, are specified, the equations can then be solved. Note that in this approximation the chemical composition of the star only enters through the parameter X.

Although the above is formally the simplest way of posing the problem it is not the easiest way to solve it numerically. The reason for this is that the boundary conditions (10.2) are rather awkward as one has to be applied at the centre of the star and the other has to be applied at the surface. There is, in fact, a single infinity of solutions of (3.75), (3.76) and (10.1) which satisfy the boundary condition at $M = 0$, each corresponding to a different value of the central density. Only one of these solutions satisfies the surface boundary condition for a star of specified mass and this must be found by a process of trial and error. It proves simpler to specify the central density of the star, ρ_c, instead of M_s. There is only one solution of (3.75), (3.76) and (10.1) which has a specified value of ρ_c and $r = 0$ at $M = 0$. This solution can be *followed* outwards until the value of M at which ρ vanishes and this gives the mass and internal structure of the star which has central density ρ_c. This procedure can then be used for a succession of values of ρ_c.

The Chandrasekhar limiting mass

When these integrations are carried out, it is found that the mass is an increasing function of the central density and the radius is a decreasing function of the central density. The more massive degenerate stars are thus smaller than the less massive ones. This is perhaps not surprising when it is realised that the total gravitational attraction which is holding the star together scales as M_s^2 while the force exerted by the pressure scales as a lower power of the mass, but one result which is at first sight very unexpected is obtained. As higher and higher values of the central density are considered, the mass of the star does not increase without limit, but it tends to a finite limiting value, known as the *Chandrasekhar limiting mass*. For masses greater than this, no models of fully degenerate stars can be constructed. I shall explain later that this should be regarded as a mathematical limit rather than a physical limit as the form of the equation of state used, (10.1), ceases to be valid before ρ_c becomes infinite. I shall also explain that the correction is one of detail rather than principle.

The value of this critical mass depends on the chemical composition of the star through the hydrogen content X. It is possible to show from (10.1), (3.75) and (3.76) that there is a very simple dependence on X. The method used is similar to that used in discussing homologous stellar models in Chapter 5. In the present case I simply give the result which can be verified to be correct. If I introduce the quantities $\bar{P}, \bar{\rho}, \bar{M}$ and $\bar{r}$, where

$$
\left.
\begin{aligned}
\bar{P} &\equiv P, \\
\bar{\rho} &= (1 + X)\rho; \\
\bar{M} &= M/(1 + X)^2, \\
\text{and} \quad \bar{r} &= r/(1 + X),
\end{aligned}
\right\}
\tag{10.3}
$$

(8.1), (3.72) and (3.73) become:

$$
\left.
\begin{aligned}
\bar{P} &= f\{\bar{\rho}\}, \\
\frac{d\bar{P}}{d\bar{M}} &= -\frac{G\bar{M}}{4\pi\bar{r}^4} \\
\text{and} \quad \frac{d\bar{r}}{d\bar{M}} &= \frac{1}{4\pi\bar{r}^2\bar{\rho}}.
\end{aligned}
\right\}
\tag{10.4}
$$

Equations (10.4) are independent of X and they can be solved to find, amongst other things, the critical value of $\bar{M}$ above which no solution of the equations is possible. When this is done, it is found that the maximum value of $\bar{M}$ is $1.44\,M_\odot$ so that the maximum mass is:

$$M_{\text{crit}} = 1.44\,(1 + X)^2 M_\odot \tag{10.5}$$

This discussion has been for the case of uniform chemical composition. It is more difficult to study non-uniform chemical composition, but a result similar to (10.5) can be obtained with an appropriate mean value of $(1 + X)^2$ on the right-hand side.

Equation (10.5) suggests a maximum possible mass of $5.76M_\odot$, when $X = 1$, but more detailed study of the problem suggests that this is a totally unrealistic value. A star as massive as $5.76M_\odot$ would pass through several of the evolutionary stages described in Chapters 6 and 9 and would burn at least a considerable fraction of its hydrogen into helium and heavier elements. In addition, a comparison of theory and observation for white dwarfs implies that most white dwarfs are unlikely to contain much hydrogen. It appears that the surface temperature of about 10^4K rises rapidly to a value of order 10^7K in the interior. At this temperature, if there was any hydrogen present, nuclear fusion reactions would occur and the star would have a much higher luminosity than that observed in white dwarfs. There is, in fact, fairly general agreement that no white dwarfs, other than those of extremely low mass, can contain any hydrogen except in the outermost layers. This implies that the maximum mass predicted by theory is about $1.44M_\odot$ (ie $X = 0$). I have already explained in Chapter 8 that if white dwarfs accrete hydrogen-rich material from a close companion a nuclear explosion may occur producing a nova.

Actual white dwarfs must be somewhat more complicated in their structure than the objects studied here. Their outermost layers will be an ideal classical gas rather than a degenerate gas and the thermal properties of the stars must be considered; in the above discussion the temperature and luminosity have not featured. When these effects are included in the theory there is no qualitative change in the results. There is a slight reduction in the predicted maximum mass and other detailed alterations.

One property of degenerate stars such as white dwarfs and neutron stars, to be discussed shortly, is that energy transport in their interiors is primarily by conduction. Because of their high interior densities, the energy possessed by the particles (electrons or neutrons) is even higher relative to that of the radiation than in ordinary stars. In addition, the particle mean free paths are increased by the degeneracy. They find it difficult to collide and give up energy because all the lower particle states are filled. The coefficient of thermal conductivity is so large that the interiors of the stars, beneath the non-degenerate outer layers, are not far from being isothermal.

For stars of mass less than the critical mass, the solution of (8.4) predicts a mass–radius relation which is shown in Table 12. Because I have not calculated the luminosity and effective temperature of these stars, I cannot place them uniquely in the HR diagram. However, for each mass a line of constant radius can be drawn

Table 12. *Mass–radius relation for fully degenerate stars containing no hydrogen*

$M_s/M_\odot$	0.2	0.4	0.6	0.8	1.0	1.2	1.4	1.44
$\log(r_\odot/r_s)$	1.68	1.81	1.90	1.99	2.10	2.24	2.57	∞

in the diagram and several such lines are shown in fig. 90. It can be seen that these constant radius lines do lie in the region of the HR diagram where the white dwarfs are found, for masses which are comparable with, but not too close to, the critical mass. Individual white dwarfs are thought to follow approximately one of these lines of constant radius as they cool down to their final dead state; there is little contraction of the star as a whole as it finally cools into invisibility.

By saying that the lines in fig 90 lie in the same region as the observed white dwarfs, I have essentially said that the actual radii of white dwarfs are similar to the radii predicted by the simple theory. In fact, with one exception which will be mentioned below, the radii of white dwarfs are not really measured directly. From the character of their radiation an estimate can be made of their surface temperature. If it is assumed that this surface temperature is not too different from the effective temperature, an estimate of the radius follows from:

$$r_s = (L_s/\pi ac T_e^4)^{1/2}, \tag{10.6}$$

and this is the method normally used to estimate the radii of white dwarfs. It is these radii which are very close to theoretical values. There seems no doubt that the radii of white dwarfs are very small.

The radii of white dwarfs can be measured with difficulty by an observation of the *gravitational red shift* of spectral lines. According to Einstein's General Theory of Relativity a photon, just like a particle, loses energy in escaping from a region of high gravitational potential. This means that the photon has lower energy when it is received than when it was emitted and this corresponds to a red shift of spectral lines. The red shift in escaping from an ordinary star is very small, but white dwarfs have a much larger value of the critical parameter $(GM_s/r_s c^2)$ and, if their masses are known, radii can be found from the red shift. The values found are in reasonable agreement with those estimated from (10.6). As well as discussing the radii of white dwarfs, it is perhaps of more interest to ask what are the masses of observed white dwarfs? Unfortunately only a small number of these are partners in easily observable binary systems, but in all cases in which a mass can be calculated it is less than the Chandrasekhar limiting mass, so that there is certainly no immediate disagreement between theory and observation. In fact most white dwarf masses deduced from observation are more like half the Chandrasekhar mass than the limiting mass.

Because there is a maximum mass for a white dwarf which realistically is more like $1.2M_\odot$ than $1.44M_\odot$, it must not be assumed that all stars which are initially more massive than this cannot become white dwarfs. I have explained in Chapter

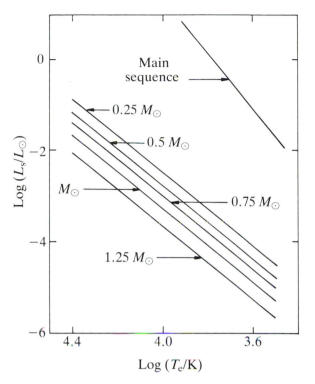

Fig. 90. HR diagram for fully degenerate stars. Such stars have a radius determined by their mass and the constant radius lines for stars of five masses are shown. A cooling white dwarf should approach the appropriate line as its temperature and luminosity decrease.

7 that mass loss from stars is common at many stages of evolution. Stars that become white dwarfs will be those which end their lives below the limiting mass without having been compressed to a higher density more appropriate to the neutron star which I shall describe shortly. The best estimate of the initial mass range of stars which become white dwarfs comes from a study of white dwarfs in galactic clusters whose age can be estimated by the method of main sequence fitting. The cluster white dwarfs must have been more massive than the stars in the cluster which are still on the main sequence. This study suggests that stars as massive as about $8M_\odot$ can become white dwarfs, so that mass loss is very important indeed.

Brown dwarfs

There is another group of stars whose pressure is due to degenerate electrons. I have explained in Chapter 5 that there is a minimum mass for a main sequence star. This depends on stellar chemical composition but is about $0.08M_\odot$. Lower mass stars never become hot enough in their interiors even to turn hydrogen into helium. They have a pre-main-sequence luminous phase, but then their interiors become degenerate and they cool. While they are visible they are given the name of brown dwarfs. They are difficult to observe even if they are very close to the Sun and it is not at present clear whether there is a large number of them. Unlike white dwarfs they have a normal chemical composition. Eventually in their dead state they, as well as white dwarfs, are called black dwarfs. If objects in the appropriate mass range do exist, there is a continuous transition from black dwarfs to large planets like Jupiter and eventually to smaller planets in which the pressure resisting gravity is that of a normal solid, rather than of a degenerate gas.

Neutron stars

I have mentioned earlier that the theory of fully degenerate stars must break down before the central density becomes infinite. In fact, according to theory, when the material is very closely packed with densities in excess of 10^{12}kg m^{-3}, the electrons combine with protons in the nuclei which are present to produce neutron rich nuclei, which would be unstable at lower densities. Because there is a decrease in the number of electrons present, the pressure of the gas is less able to resist gravity and the star contracts. This in turn leads to further production of neutrons. This is the main factor which makes the actual maximum white dwarf mass less than the Chandrasekhar mass. Eventually at densities in excess of about 10^{15}kg m^{-3} almost all of the material is in the form of neutrons, with just a very small admixture of electrons, protons and heavier nuclei. At these densities the neutrons form a degenerate gas rather than an ideal classical gas and the star is then known as a *neutron star*. As explained in Chapters 8 and 9, neutron stars are present in some X-ray binaries and are the end product of some supernova explosions.

There are three separate reasons why the equations for the structure of neutron stars differ from the equations for white dwarfs. The first is the simple one that in a neutron star the particles which provide the mass are also the particles that provide the pressure, whereas in white dwarfs essentially all of the pressure is provided by the electrons, while the mass is almost entirely atomic nuclei. The next effect is that influences of both special relativity and general relativity must be taken into account. Special relativity states that all energy has a mass equivalent and in neutron stars, but not in white dwarfs, the mass equivalent of the kinetic energy is important. General relativity introduces a special radius, the *Schwarzschild radius*,

$$R_{\text{sch}} = 2GM/c^2,\tag{10.7}$$

which a spherical body of mass M cannot enter without collapsing to a point. The relativistic terms change the equation of hydrostatic support to

$$\frac{dP}{dr} = -G\left(M + \frac{4\pi Pr^3}{c^2}\right)\left(\rho + \frac{P}{c^2}\right)\Big/r^2\left(1 - \frac{2GM}{rc^2}\right).\tag{10.8}$$

The three changes in (10.8) compared with (3.4) all increase the gravitational

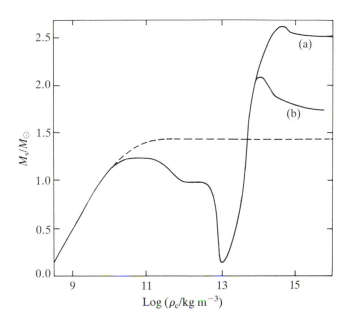

Fig. 91. The relation between mass and central density for white dwarfs and neutron stars. Curves (a) and (b) correspond to different assumptions about the equation of state at high densities. The dashed curve gives the Chandrasekhar relation for white dwarfs.

attraction and, if there were no other effects, the maximum mass of a neutron star would be less than that of a white dwarf. It is known as the Oppenheimer–Volkoff mass and its value is about $0.7M_\odot$.

There is a third effect which is that the pressure is not given by a straightforward equivalent of (4.49) and (4.51). When neutrons are packed very close together at a density in excess of that of an atomic nucleus, they repel one another and this increases the pressure over the value given by the Pauli exclusion principle alone. Experiments cannot provide the equation of state at such densities and there are quite large differences between theoretical estimates. They all agree that there must be a maximum mass for a neutron star but the value remains uncertain. Most estimates place it in the range $2M_\odot$–$2.5M_\odot$. The mass/central density relations for two theoretical equations of state are shown in fig. 91, which also shows the white dwarf relationship. Note that the maximum mass of a neutron star does not occur at the highest possible central density. It is not, however, believed that any star can die with a finite radius if its final mass exceeds the maximum neutron star mass. It must then become a black hole which I shall describe shortly. As in the case of a white dwarf, the initial mass of a star which becomes a neutron star may be very much larger than its final mass. I have already mentioned the neutron star as the end product of a Type II supernova explosion in Chapter 9. It is possible that all neutron stars are produced in supernovae but it is also possible that a white dwarf, particularly in a close binary system of the type described in Chapter 8, could accrete matter until its mass exceeded the maximum white dwarf mass and then collapse to become a neutron star. Neutron stars have been discovered in two ways. As explained in Chapter 8, in X-ray emitting close binary systems a visible star is found to be orbiting an invisible companion whose mass is too great to be a white dwarf. In addition there are *pulsars* which I now describe.

Pulsars

Although the possible existence of neutron stars was discussed in the 1930s, it was only in the 1960s that any observations were made which suggested their existence. I have mentioned in Chapter 9 the class of object known as pulsars, which emit periodic bursts of radio waves with a very short period. The radiation from a pulsar may have the character shown in fig. 92; there are significant variations in the character of the emission from pulse to pulse but the time between pulses is constant apart from a very small effect to be mentioned shortly. In the earliest pulsars discovered the inter-pulse period was of order 1s and the pulse length was about 30ms. The character of the pulses shows that the region from which radiation is emitted is very small. If radiation is emitted simultaneously from a region of finite extent, the arrival time of the radiation at an observer varies because of the different distances which radiation from different parts of the source has to travel (see fig. 93). For a pulse to last no longer than about 30 ms the emitting region should have a linear extent no greater than about 10^7 m and probably much smaller. In a minority of pulsars, including that in the Crab nebula, there are two pulses of differing intensity in each period.

If the emission is coming from a reasonable fraction of the object concerned, only white dwarfs and neutron stars seem small enough to be pulsars. Initially three suggestions were made for the possible source of the periodicity: the rotation period of a single object, an orbital period of a binary system or a period of radial pulsation, as in cepheid variable stars. Only very small objects can have any of these periods as small as 1s, if no velocities are to exceed the velocity of light, and once a pulsar in the Crab nebula was discovered with a period as small as 30ms, it became accepted that pulsars were neutron stars and not white dwarfs. More recently pulsars with periods of a few ms have been found. For them even a neutron star must be involved in motions at a significant fraction of the speed of

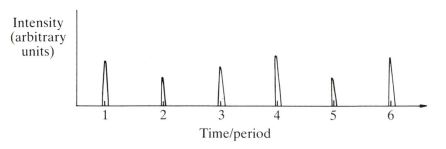

Fig. 92.   The radiation from a typical pulsar. The properties of individual pulses can be very variable but the time between pulses is almost precisely constant.

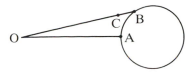

Fig. 93.   Light travel time from an extended source. Light from two points A and B takes different times to reach the observer O. Even if a sharp pulse is emitted simultaneously over the source between A and B, it must be spread out on arrival by the time it takes light to travel from B to C.

light. The discovery of the Crab nebula pulsar was the first clear association of a neutron star with the remnant of a supernova explosion.

After pulsars had been studied for some time it was found that their periods were slowly increasing. This led to the accepted model of pulsars. They are strongly magnetised and rapidly rotating neutron stars in which the axis of rotation and that of the magnetic field do not coincide. Such an object must radiate electromagnetic waves. The energy that is radiated comes from the rotational kinetic energy of the pulsar and that is why the period gradually increases. The total energy radiated is significantly greater than that in the radio pulse, which led to the discovery of pulsars, and it emerges in many different regions of the electromagnetic spectrum. In some millisecond pulsars the story may be a bit more complicated. Here accretion of matter from a companion star, of the type which I have described in Chapter 8, may cause the neutron star to spin up and its period to decrease. The nature of the pulsed radiation is still not completely clear. It is believed that it comes from very high energy particles accelerated in a magnetic field and that as the pulsar rotates the emission reaches the observer in the same manner as a rotating searchlight beam.

A pulsar in a binary system has been responsible for the almost certain verification of a prediction of general relativity. The theory predicts that suf-ficiently asymmetric rotating bodies radiate waves, known as gravitational waves. Because gravitation is so weak compared to electromagnetism, gravitational waves are only significant for objects which are not much larger than their Schwarzschild radius and in which velocities which are not much less than the speed of light are present. This is the case in the binary pulsar in which a pulsar has a second neutron star as its companion. It is predicted that radiation of gravi-tational waves should produce rapid changes in the orbital properties of the system and these have been observed.† Astronomers are attempting to detect gravitational waves by laboratory experiments. Although progress has been made, no existing experiment yet has the accuracy required to identify cosmic sources of gravitational waves with their expected intensity.

---

† This work was recognised by the award of the 1993 Nobel Prize for Physics to R. A. Hulse and J. H. Taylor.

Black holes

With the discussion which I have given above of the structure of white dwarfs and neutron stars, it seems that we have an approximate description of the final stage in the life history of stars whose masses then are less than about $2M_\odot$. These stars can pass through a sequence of phases of nuclear evolution, their central temperatures can reach a maximum value and then fall and the stars can die. It is quite probable that their lives are more violent and exciting than this description suggests. As I have explained, mass loss is very important in stellar evolution and the final mass of many stars is very much smaller than the initial mass. Indeed neutron stars may only be formed as a direct result of a stellar explosion. There does, however, remain the probability that some stars have a final mass greater than the maximum neutron star mass. What becomes of them?

For a long time astronomers avoided this question and also disregarded the possible existence of neutron stars by assuming that all stars would lose sufficient mass to become white dwarfs. The discovery of neutron stars in the 1960s also led to the acceptance of the possibility that, within the laws of physics as they are at present understood, there is nothing to prevent a more massive star collapsing to a state of zero radius and infinite density. Such an object is called a black hole. As the size of such an object approaches its Schwarzschild radius, its radiation is shifted to the red by the gravitational redshift to such an extent that it becomes effectively invisible. Once an object is inside its Schwarzschild radius, it can no longer communicate with its surroundings except that it still exerts its gravitational influence on surrounding matter; an electrically charged black hole would also exert an electrical force.

As I have already mentioned in Chapter 8, the place to look for stellar mass black holes is in close binary systems where a visible star may be found to be orbiting about an invisible companion which is too massive to be a neutron star. It is believed, as a result of such observations, that a black hole must be present in the X-ray emitting binary Cygnus X-1. The presence of much more massive black holes than can result from the collapse of a single star is suspected in the nuclei of some galaxies and in quasars. Because isolated black holes do not radiate, it is possible that there could be a considerable amount of hidden matter in the Universe in the form of black holes.

I have stated earlier that the actual value of the maximum mass for neutron stars is uncertain, but that the existence of a maximum mass seems clear. It can be shown that there must be such a maximum mass if the force of gravitation is always attractive, however close together particles are, and if the postulate of the special theory of relativity that no energy can be propagated with a velocity greater than that of light is true. However, the peculiar nature of gravitational collapse has led some people to ask whether it is not possible that some of the presently accepted laws of physics go seriously wrong at very high densities.

If the discussion in this chapter is correct, a vast majority of all stars will eventually become black dwarfs. A smaller number will become neutron stars and an even smaller number black holes. The disparity in numbers might be less than appears from our knowledge of the distribution of stellar masses in our Galaxy today because there have been suggestions that the earliest generations of stars were predominantly massive stars. However, this remains very uncertain. There is finally a possibility that some stars may not become a black dwarf, neutron star or black hole. Some calculations of stellar explosions have suggested that they might occasionally be so efficient that no remnant at all is left behind. In that case the whole of the mass of a star would become available to be incorporated in later generations of stars.

Summary of Chapter 10

In this chapter I have discussed the final stages of stellar evolution. If the electrons in a star's central regions become degenerate, it is possible that its central temperature may stop increasing and begin to fall. After this, no further significant nuclear fusion reactions occur and the star gradually cools and *dies*. It is believed that white dwarfs are stars of this type. In addition very low mass stars become degenerate and cool down without any nuclear reactions occurring. They are known as brown dwarfs. Brown dwarfs and white dwarfs both eventually become black dwarfs. If such a star is sufficiently massive at the end of its life ($\geqslant 1.4 M_\odot$), the degeneracy pressure of the electrons eventually proves insufficient to resist the attractive gravitational force and the star contracts further. At extremely high densities protons and electrons combine until the star consists almost entirely of neutrons. Neutron stars have been found as pulsars and also as X-ray emitting objects in some close binary systems. There also appears to be a maximum possible mass for a neutron star. The value of this mass is not well known, but it is thought to be rather greater than the maximum mass for a white dwarf. Above this mass, if the presently accepted laws of physics are correct, objects enter into a rapid gravitational collapse which cannot be halted and become black holes.

11

Concluding remarks

In this book I have described the methods used in the theoretical study of stellar structure and evolution and I have discussed many of the results obtained. I have tried to discuss the present state of a developing subject and to mention the main uncertainties. As I have stressed, particularly at the end of Chapter 9, some of the detailed theoretical ideas may prove to be wrong, but it is confidently expected that the broad outline of the subject as presented in Chapters 3–5 is correct. In this chapter I discuss further some of the points where important uncertainties remain.

In the first place it is important to realise that, although this book has been written by a theoretical astrophysicist, who has a particular interest in obtaining a theoretical understanding of the subject, ultimately all of the theoretical work must be related to observations. This has a twofold implication. The theoretical worker must keep the observational results in mind and there is a continuing need for new observations. The subject depends considerably on some of the less glamorous parts of observational astronomy. In these days of quasars, pulsars and the cosmic microwave radiation, the work of measuring parallaxes and proper motions and studying the orbits of binary star systems is often regarded as being very humdrum. However, it is vitally important in supplementing the information possessed about such things as masses, radii and absolute magnitudes. As has been mentioned in Chapter 2, the amount of reliable information which we have about some of these quantities is very small, but further results can be obtained by patient observations. The profile of the subject has been raised by the satellite HIPPARCOS, specially designed to study parallaxes and proper motions.

Since the first edition of this book appeared observations have provided much very important information about stellar evolution. In 1970 there were very few significant observations relating to star formation. It was possible then for theoreticians to study pre-main-sequence stellar evolution on the basis of the contraction of a spherical protostar, not because it was regarded that this was obviously correct but because there were no observations indicating how the simplest model needed to be modified. Now observations in the infrared and

millimetre regions of the spectrum are providing very complex information about
sites of star formation and are indicating that protostars have a complex geometry,
including in some cases both accretion from a disk and bipolar outflows along the
axis of rotation. Given that many, if not most, stars are members of binary or
multiple systems, there is considerable interest in trying to determine which
factors lead to binary stars and which to planetary systems. The overall theoretical
problem of star formation, including the influence of both magnetic fields and
rotation, is likely to be an active research area for many years.

As has been explained in Chapter 7, extension of observations into the
ultraviolet region of the spectrum has led to the realisation that mass loss is very
important at many stages of stellar evolution. It is now clear that massive stars end
their lives with a very much smaller mass than when they start. Considerable
uncertainties remain about precisely how much mass a star can be expected to lose
and this influences its expected end state, black dwarf, neutron star or black hole.
This, in turn, through the chemical composition of the ejected mass, influences
ideas about the origin of the chemical elements and the chemical evolution of
galaxies. It has also become clear that many low mass stars have active outer
regions like the Sun and that many have stellar winds, which are much stronger
than the solar wind. The evolution of these low mass stars cannot be fully
understood in terms of isolated spherical stars because both rotation and magnetic
fields are obviously important.

The single observation which did most to focus the attention of theoreticians in
recent years was the explosion of supernova 1987A. Theoretical ideas about
supernova explosions were put to the test in a manner which was not possible
before. Some of the ideas passed the test very successfully. The detection of
neutrinos showed that most of the energy liberated *does* come out in the form of
neutrinos and it was confirmed that a significant amount of ^{56}Ni was produced and
that its decay formed the declining light curve. The supernova, however, raised as
many questions as it answered. The explosion of a blue supergiant was unexpected
and the deduced mass of the star was such that the theoretical model did not lead
to an explosion. Furthermore important departures from spherical symmetry
were seen in the ejected matter. All of the observations have stimulated increased
theoretical studies of massive supernovae.

The basic physics of stellar interiors is believed to be reasonably well-
understood. There have been some important recent developments in calculations of stellar opacities at low temperatures of order 10^6K or less. Although
these do not make a very large difference to the calculated overall properties of
stars, they appear to be just what is required to explain the existence of some types
of variable star. As the stars pulsate there are changes in the value of the opacity
which modulate the ease with which radiation can escape from the stars.
Otherwise the main uncertainty continues to be the theory of well-developed
convection. There have been no improvements on the mixing length theory. Even
if this gives a qualitatively good description of energy transport by convection,
there is no way in which the appropriate value of the mixing length can be
determined other than by demanding that theory and observation be brought into

agreement. Unfortunately solutions to difficult non-linear problems such as that of large amplitude convection are not readily obtained.

The other major worry about early stellar evolution is the continuing discrepancy between the predictions and results of solar neutrino experiments. It may be that this discrepancy has nothing to do with astronomy and that it is entirely due to the behaviour of the neutrinos emitted in the hydrogen-burning reactions. It will not be possible to be really happy with the foundations of the theory of stellar structure and stellar evolution until this matter is resolved, although I continue to believe that the theory has so many other successes that the ultimate solution will leave it essentially intact.

Another problem in which precise information about the evolution of low mass stars is required is the determination of the ages of globular clusters which I have discussed briefly in Chapter 6. As I mentioned there, most workers find an age of order 1.5×10^{10} years for globular star clusters, or at least for the oldest ones. The globular star clusters are the oldest known component in the Galaxy and their age probably provides a good estimate of when the Galaxy started to form. Ideas about galaxy formation and observations of quasars suggest that the Universe must have been at least about 10^9 years old before the Galaxy formed. If that is the case we require the Universe to be of order 1.6×10^{10} years old. Cosmology also provides an estimate of the age of the Universe, which is still not very precise principally because of continuing disagreements about the value of Hubble's constant. The age of the Universe determined by the globular clusters is larger than most, but not all, estimates of the age provided by cosmology. There is therefore a particular interest in asking whether uncertainties in ideas about stellar evolution might allow the globular clusters to be younger. Changes in the assumed value of the mixing length have some effect. Another idea is that, whereas it is usually assumed that the chemical composition of the outer regions of stars is uniform, helium settles relative to hydrogen in the star's gravitational field. There is some evidence for such segregation of elements in one group of stars known as peculiar A stars. Helium diffusion has some effect on the estimated ages of globular clusters but not a very large effect.

Increasingly the problems which are being studied in stellar evolution are ones which could not be tackled before the development of very fast computers with large storage capacity. Initially most calculations were concerned with single spherical stars in those stages of evolution in which it was possible to ignore the time derivatives in the equations of the stellar structure. It then became possible to follow the true time-dependent behaviour of such spherical stars, except possibly in the most rapid stages of evolution such as the helium flash and the dynamics of a supernova explosion. Now it is becoming possible to attempt serious discussions of non-spherical stars, with particular emphasis being placed on close binary stars because, as mentioned in Chapter 8, many of the most dramatic events in stellar evolution involve close binaries. The discussion of the fine detail of the structure of such stars, including for example the effects of the irradiation of the secondary star by the primary, are, however, still in a primitive stage.

There are many interesting problems relating to single non-spherical stars, where the departure from sphericity is caused by rapid rotation or by a strong magnetic field. One question which has been asked in the case of the Sun is whether the surface rotation and magnetic field of a star are necessarily characteristic of the values in the deep interior. Early attempts to resolve the solar neutrino problem included suggestions that the Sun possessed either a very rapidly rotating core or a very strong internal magnetic field. In either case the central temperature of the Sun would be reduced relative to its standard value and this would reduce the flux of neutrinos capable of being detected by ^{37}Cl. Both of these ideas have been rejected. Both of them would lead to a departure from sphericity of the solar surface which is larger than observed. In addition the magnetic field would be unstable while a very strong concentration of angular momentum to the solar centre is incompatible with the properties of solar oscillations, which have been mentioned briefly in Chapter 6.

There is no static equilibrium state of a star which is rotating or which contains a magnetic field. I have mentioned this briefly in Chapter 6, when I mentioned that it used to be thought that stars were kept well mixed by meridional circulation. Although it is now thought that such mixing is unimportant in most stars, it should be significant in some rapidly rotating and strongly magnetic stars. As well as mixing the stellar material, such circulation has an effect on the rotation law of the star. The viscosity of stellar material is not very effective in reducing relative velocities between elements of stellar material in a star's lifetime. This means that each element tends to retain its own angular momentum about the star's axis of rotation as the law of conservation of angular momentum applies to each element individually. As meridional circulation moves an element around, it moves into regions previously occupied by elements with a differing angular momentum per unit mass and this modifies the distribution of angular momentum and angular velocity through the star. Similar effects are related to the conservation of magnetic flux. If a differentially rotating star is also threaded by a magnetic field, the rotation tends to wind up the field which then reacts back on the rotation. Even quite a weak magnetic field can restrict the degree of differential rotation in a star. The above effects apply in stellar radiative zones. In convection zones, the properties of the convection are affected by both rotation and magnetic fields, and it is a complex interaction between rotation, magnetic fields and convection which produces the surface activity in the Sun and other late type stars.

The detailed discussion in this book has been concerned with single spherical stars. Although there continue to be uncertainties and unsolved problems concerning their structure and evolution, very substantial progress has been made. Observations of the surface properties of stars have been combined with a few basic physical laws and many measured and calculated properties of atoms and nuclei to give what is believed to be reasonably accurate information about the unobserved interiors of stars. In most cases these interiors will be unobservable for ever. The solar neutrino experiment and solar oscillations probe the internal structure of the Sun and there are hopes of extending the latter technique to stars. Supernova explosions and less violent phases of mass loss can expose evolved cores of stars and in other cases the products of nuclear reactions in a star are mixed to its surface. These observations provide some check on the theoretical ideas. The subject of stellar evolution is now a mature subject in which the easy problems have been solved but in which many very interesting and difficult problems remain. In the first edition in 1970, I concluded by commenting on the need for a resolution of the solar neutrino problem and by expressing my own belief that it would be resolved in such a way as to leave the theory of stellar structure on good foundations. I will end by expressing the same view almost twenty five years later.

Appendix 1

Thermodynamic equilibrium

If a physical system is isolated and left alone for a sufficiently long time, it settles down into what is known as a state of thermodynamic equilibrium. In thermodynamic equilibrium the overall properties of the system do not vary from point to point and do not change with time. Individual particles of the system are in motion and do have changing properties. For example, electrons may be being removed from and attached to atoms. There is, however, a statistically steady state in which any process and its inverse occur equally frequently. Thus, in the example mentioned above, the number of atoms ionised per unit time is equal to the number of recombinations. Because the properties of a system do not vary from point to point when it has reached thermodynamic equilibrium, all parts of it have the same temperature.

If two such isolated systems are brought into contact, heat will flow from one to the other until they reach the same state of thermodynamic equilibrium and hence the same temperature. In thermodynamic equilibrium all of the physical properties of the system (such as pressure, internal energy, specific heat) can be calculated in terms of its density, temperature and chemical composition alone. In nature a true state of thermodynamic equilibrium may be approached closely but never quite reached.

In thermodynamic equilibrium the intensity of radiation is given by the Planck function

$$I_\nu = B_\nu(T) \equiv \frac{2h\nu^3/c^2}{\exp(h\nu/kT) - 1}. \tag{A1.1}$$

It should be noted that the Planck function is determined by the temperature alone and does not depend on the density and chemical composition of the material. A particular consequence of thermodynamic equilibrium is Kirchhoff's law

$$j_\nu = \kappa_\nu B_\nu, \tag{A1.2}$$

where j_ν and κ_ν are emission and absorption coefficients for radiation of frequency ν.

Provided corrections due to quantum mechanics and relativity are slight (and this is often true), any species of particle possesses a Maxwellian velocity distribution

$$f(u, v, w) = n\left(\frac{m}{2\pi kT}\right)^{3/2} \exp\left(-m(u^2 + v^2 + w^2)/2kT\right), \qquad (A1.3)$$

where n is the number of the particles in unit volume, m the particle mass, u, v, w the three components of velocity of the particles and the number of particles in unit volume with velocities between (u, v, w) and $(u + \delta u, v + \delta v, w + \delta w)$ is $f\,\delta u\,\delta v\,\delta w$. The corresponding distribution function for a gas in which quantum effects are important is stated in Appendix 3.

In any particular atom or ion, electrons will be arranged in various energy levels. If two such levels have energies E_r and E_s, in the thermodynamic equilibrium the numbers n_r, n_s of atoms in the two states will obey the Boltzmann law

$$\frac{n_r}{n_s} = \frac{\exp\left(-E_r/kT\right)}{\exp\left(-E_s/kT\right)}. \qquad (A1.4)$$

Finally an atom can exist in several states of ionisation and the number of atoms per unit volume in two successive states of ionisation, n_i,† n_{i+1} is related to the electron density n_e by Saha's equation

$$\frac{n_{i+1}n_e}{n_i} = \left(\frac{2\pi mkT}{h^2}\right)^{3/2} \frac{2B_{i+1}}{B_i} \exp\left(-I_i/kT\right), \qquad (A1.5)$$

where I_i is the energy required to remove one electron from the atom in the ith state of ionisation and B_i, B_{i+1}, which are called the partition functions for the two states, depend on the electron energy levels in the two ions and the temperature. The chemical composition of the medium enters into Saha's equation because the electrons entering into n_e can be provided by ionisation of all of the elements present.

In the deep interior of a star departures from thermodynamic equilibrium are slight but, as the surface of the star is approached, (A1.1) at least must cease to be true. For one reason, the radiation primarily flows outward and in addition the distribution of radiation with frequency ceases to have the Planck form. If collisions between particles are sufficiently rapid, (A1.2)–(A1.5) may still continue to be valid for some quantity T, which is not a true thermodynamic temperature but which is known as the kinetic temperature. If this is true the system is said to be in a state of *local thermodynamic equilibrium*. When population of the states of excitation and ionisation follow (A1.4) and (A1.5), interpretation of properties of spectral lines in terms of element abundances and

† The quantities in (A1.4) should now be written more correctly n_{ir}, n_{is}.

other stellar properties such as rotational velocity is relatively straightforward. Very near the surfaces of stars it seems certain that the approximations of local thermodynamic equilibrium break down and it is then much more difficult to convert observed line strengths into abundances. This is particularly true of regions where emission lines are formed and that is why stellar emission lines rarely yield reliable abundances. I do not discuss stellar atmospheres and the determination of stellar chemical composition in this book. In the interstellar medium conditions are very far from thermodynamic equilibrium.

In the above discussion it has been assumed that the chemical composition of the material can be specified but, in fact, that is not strictly true. Nuclear reactions gradually convert nuclei from one form to another. Usually nuclear reactions occur so slowly that the assumption of unchanging chemical composition is justified in discussing the approach to thermodynamic equilibrium. In conditions of sufficiently high temperature and density nuclear reactions may occur so rapidly that an approach to nuclear statistical equilibrium takes place. If this happens the chemical composition of the material can no longer be specified in advance but it must be determined by using a set of Saha-type equations, which couple the densities of the free protons and neutrons and the complex nuclei. In complete thermodynamic equilibrium the properties of a system are determined by its density and temperature and by certain conserved quantities, such as, in this example, the total number of neutrons and protons. It is possible that a close approximation to complete thermodynamic equilibrium has occurred in some stars immediately before a supernova explosion and also in the very early stages of the expanding Universe. In the latter case many elementary particles in addition to neutrons, protons and electrons partake in the equilibrium.

Appendix 2

The equation of radiative transfer

The intensity of radiation $I_\nu(\mathbf{x}, \mathbf{k})$ at a point $\mathbf{x}$ in a star in equilibrium is defined by

$$\delta E = I_\nu(\mathbf{x}, \mathbf{k})\delta S \, \delta\omega \, \delta\nu \, \delta t, \tag{A2.1}$$

where δE is the energy crossing an area δS perpendicular to a unit vector $\mathbf{k}$ in solid angle $\delta\omega$ about $\mathbf{k}$, in frequency range $\delta\nu$ about ν, in time δt (see fig. 94). This intensity is almost isotropic (independent of direction $\mathbf{k}$) and almost equal to the Planck function B_ν. I can therefore write

$$I_\nu(\mathbf{x}, \mathbf{k}) = B_\nu(\mathbf{x}) + \delta_\nu(\mathbf{x}, \mathbf{k}), \tag{A2.2}$$

where $\delta_\nu(\ll B_\nu)$ is the slight departure from isotropy that leads to a net flow of radiation through the star.

As the radiation moves a distance δs along $\mathbf{k}$, I_ν changes for three reasons.

(i) Radiation is absorbed.

I define a mass absorption coefficient $\kappa_\nu(\mathbf{x})$ by

$$(\delta I_\nu)_{\text{abs}} = -\kappa_\nu \rho I_\nu \, \delta s. \tag{A2.3}$$

κ_ν is independent of $\mathbf{k}$ so long as the absorbing atoms or ions have a random distribution of orientations.

(ii) Radiation is scattered, both out of and into direction $\mathbf{k}$.

If I introduce a mass scattering coefficient $\sigma_\nu(\mathbf{x})$, I can write

$$(\delta I_\nu)_{\text{scatt}} = -\sigma_\nu \rho I_\nu \, \delta s + \sigma_\nu \rho [\int I_\nu(\mathbf{k}')p(\mathbf{k}, \mathbf{k}') \, d\omega'] \, \delta s. \tag{A2.4}$$

$p(\mathbf{k}, \mathbf{k}')$ is the probability that radiation scattered from direction $\mathbf{k}'$ goes into direction $\mathbf{k}$ and $d\omega'$ is an element of solid angle about $\mathbf{k}'$. In (A2.4) I have assumed that σ_ν is isotropic, which will also be true for randomly orientated scatterers, and that there is no change of frequency on scattering. This is true to a very close

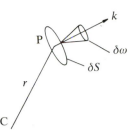

Fig. 94.   Radiation flow at a point P in a star distant r from the centre C.

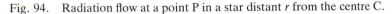

approximation for the main source of scattering in stellar interiors, scattering by free electrons.

(*iii*) *Radiation is emitted.*

I define $j_\nu(\mathbf{x}, \mathbf{k})$ to be the energy emitted per unit mass, per unit frequency range, per unit solid angle about $\mathbf{k}$, per second. Then

$$(\delta I_\nu)_{\text{em}} = j_\nu \rho \delta s. \tag{A2.5}$$

With this definition j_ν is not isotropic. However it has two parts:

(a) spontaneous emission, which is isotropic;
(b) stimulated emission, which follows absorption and is in the same direction as the absorbed radiation.

In conditions close to thermodynamic equilibrium, it can be shown that

$$(j_\nu)_{\text{stim}} = \kappa_\nu I_\nu \exp{(-h\nu/kT)}. \tag{A2.6}$$

This stimulated emission just looks like negative absorption. It is convenient to replace j_ν and κ_ν by j'_ν and κ'_ν, which are the coefficient of spontaneous emission and the absorption coefficient corrected for stimulated emission. Thus

$$j'_\nu = j_\nu - (j_\nu)_{\text{stim}}. \tag{A2.7}$$

and

$$\kappa'_\nu = \kappa_\nu[1 - \exp{(-h\nu/kT}], \tag{A2.8}$$

where j'_ν and κ'_ν are both isotropic.

The total change in I_ν can now be written:

$$\delta I_\nu = j'_\nu \rho \delta s - \kappa'_\nu \rho I_\nu \delta s - \sigma_\nu \rho I_\nu \delta s + \sigma_\nu \rho [\int I_\nu(\mathbf{k}') p(\mathbf{k}, \mathbf{k}') \, d\omega'] \delta s. \tag{A2.9}$$

In thermodynamic equilibrium (see Appendix 1) $I_\nu = B_\nu$ and j'_ν and κ'_ν satisfy Kirchhoff's law $j'_\nu = \kappa'_\nu B_\nu$. Since conditions in stellar interiors are close to thermodynamic equilibrium, we can write

$$j'_\nu = \kappa'_\nu B_\nu + \delta'_\nu(\mathbf{x}), \tag{A2.10}$$

where

$$\delta'_\nu \ll \kappa'_\nu B_\nu. \tag{A2.11}$$

I now substitute (A2.2) and (A2.10) for I_ν and j'_ν in (A2.9) and at the same time divide by δs and take the limit $\delta s \to 0$ to obtain a differential equation., After cancellation of some terms and one approximation, I obtain

$$\frac{dB_\nu}{ds} = \delta'_\nu \rho - (\kappa'_\nu + \sigma_\nu)\rho\delta_\nu + \sigma_\nu\rho \int \delta_\nu(\mathbf{k}')p(\mathbf{k}, \mathbf{k}')\, d\omega'. \tag{A2.12}$$

To obtain this result I have used the fact that B_ν is isotropic and that $\int p(\mathbf{k}, \mathbf{k}')\, d\omega'$ = 1. I have also neglected $d\delta_\nu/ds$. This approximation is justified because in thermodynamic equilibrium, when $\delta_\nu \equiv \delta'_\nu \equiv 0$, $dB_\nu/ds = 0$. This means that dB_ν/ds is small for small departures from thermodynamic equilibrium and $d\delta_\nu/ds$ must be even less important.

I now make a further approximation which is essentially correct for important scattering processes in stellar interiors, which is that there is symmetry between forward and backward scattering. This means $p(\mathbf{k}, \mathbf{k}') = p(\mathbf{k}, -\mathbf{k}')$. If this is true, it can be verified that a solution of (A2.12) is

$$\delta_\nu = -\frac{1}{(\kappa'_\nu + \sigma_\nu)\rho}\frac{dB_\nu}{ds} + \frac{\delta'_\nu}{\kappa'_\nu}. \tag{A2.13}$$

Given that δ'_ν, κ'_ν and σ_ν are isotropic, the only term which does not obviously cancel with another term when (A2.13) is inserted into (A2.12) is

$$-\frac{\sigma_\nu}{(\kappa'_\nu + \sigma_\nu)}\int \left(\frac{dB_\nu}{ds}\right)_{\mathbf{k}'} p(\mathbf{k}, \mathbf{k}')\, d\omega'. \tag{A2.14}$$

However (A2.14) does vanish if $p(\mathbf{k}, \mathbf{k}') = p(\mathbf{k}, -\mathbf{k}')$, because $(dB_\nu/ds)_{\mathbf{k}'} = -(dB_\nu/ds)_{-\mathbf{k}'}$ and the integral over any one hemisphere is equal and opposite to the integral over the opposite hemisphere.

Thus I have

$$I_\nu(\mathbf{x}, \mathbf{k}) = B_\nu(\mathbf{x}) - \frac{1}{(\kappa'_\nu + \sigma_\nu)\rho}\frac{dB_\nu}{ds} + \frac{\delta'_\nu(\mathbf{x})}{\kappa'_\nu}. \tag{A2.15}$$

Only the second term on the right hand side of (A2.15) is not isotropic and contributes to the net flow of radiation. In spherical symmetry (fig. 95)

$$\frac{dB_\nu}{ds} = \cos\theta\frac{dB_\nu}{dr} = \cos\theta\frac{dB_\nu}{dT}\frac{dT}{dr}. \tag{A2.16}$$

Also the net flow of radiation, at all frequencies, across a spherical surface of radius r, is

$$L(r) = \int_0^\infty L_\nu\, d\nu = \int_0^\infty \left[4\pi r^2 \int I_\nu(r, \mathbf{k})\cos\theta\, d\omega\right] d\nu, \tag{A2.17}$$

where the $I_\nu \cos\theta$ inside the brackets is the part of the radiation moving in

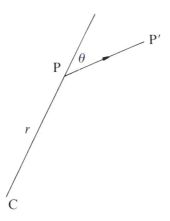

Fig. 95.   In a spherical star the value of the Planck function B_ν only changes because r changes. As a result the rate of change of B_ν in the direction PP′ is $\cos\theta$ times the radial rate of change.

direction **k**, which makes a contribution to the flow in the radial direction and the integral dω is over all directions **k**. Then using d$\omega = \sin\theta$ dθ dφ and $\int \cos^2\theta$ d$\omega = 4\pi/3$ and the expressions (A2.15) and (A2.16) I obtain

$$L(r) = -\frac{16\pi^2 r^2}{3\rho}\frac{dT}{dr}\int_0^\infty \frac{dB_\nu}{dT}\frac{d\nu}{(\kappa'_\nu + \sigma_\nu)}. \tag{A2.18}$$

It is now useful to write (A2.18)

$$L(r) = -\frac{16\pi^2 r^2}{3\rho}\frac{dT}{dr}\int_0^\infty \frac{dB_\nu}{dT}d\nu\left[\left.\int_0^\infty \frac{dB_\nu}{dT}\frac{d\nu}{(\kappa'_\nu + \sigma_\nu)}\right/\int_0^\infty \frac{dB_\nu}{dT}d\nu\right]. \tag{A2.19}$$

A well-known property of the Planck function is

$$\int_0^\infty B_\nu\, d\nu = acT^4/4\pi \tag{A2.20}$$

so that

$$\int_0^\infty \frac{dB_\nu}{dT}\, d\nu = acT^3/\pi. \tag{A2.21}$$

Then (A2.19) can be written

$$L(r) = -\frac{16\pi acr^2 T^3}{3\kappa\rho}\frac{dT}{dr}, \tag{A2.22}$$

provided that I define κ by

$$\frac{1}{\kappa} = \left.\int_0^\infty \frac{dB_\nu}{dT}\frac{d\nu}{(\kappa'_\nu + \sigma_\nu)}\right/\int_0^\infty \frac{dB_\nu}{dT}\, d\nu. \tag{A2.23}$$

κ is known as the Rosseland mean absorption coefficient or the opacity. Equation (A2.22) can be rewritten

$$\frac{dT}{dr} = -\frac{3\kappa L\rho}{16\pi acr^2 T^3}. \tag{A2.24}$$

which is the form in which I normally use it.

Energy transport by conduction can also be included in this equation if κ is defined by

$$\frac{1}{\kappa} = \frac{1}{\kappa_R} + \frac{3\rho\lambda_{cond}}{4acT^3}. \tag{A2.25}$$

where κ_R is given by (A2.23) and λ_{cond} is the coefficient of thermal conductivity so that the conductive luminosity is

$$L_{cond} = -4\pi r^2\lambda_{cond}\, dT/dr.$$

Near to the surface of a star I_ν deviates greatly from B_ν and a different approach must be taken to the solution of (A2.9) to obtain the properties of the emergent spectrum from the star. This treatment is outside the scope of the present book.

Appendix 3

The pressure of a degenerate gas

Electrons, protons and neutrons are all particles which are known as *fermions*. Such particles obey a quantum mechanical principle known as the *Pauli exclusion principle*, which states that no two identical fermions can occupy a single quantum state. If I consider free particles not bound to atoms or molecules, I can specify the state of a particle by the values of the three components of its position and the three components of its momentum. The six-dimensional space of position and momentum is named *phase space* and quantum mechanical calculations further indicate that I must associate one particle state with each volume h^3 in phase space. Fermions also possess an intrinsic property called spin which for electrons, protons and neutrons can have two values. Taking account of the spin state, the Pauli exclusion principle allows at most two of any one of these particles to be present in any volume h^3 in phase space.

Particles usually have a Maxwellian distribution of velocity given by (A1.3). This can be written in terms of momentum and I can ask whether the Maxwellian distribution violates the Pauli exclusion principle. It is convenient to discuss how many particles in a volume V in ordinary space can have a magnitude of momentum between p and $p + \delta p$. Using spherical polar coordinates in momentum space, a spherical shell has volume $4\pi p^2 \delta p$. The corresponding volume in phase space is $4\pi V p^2 \delta p$, the number of quantum states is $(4\pi V p^2 / h^3) \delta p$ and the number of fermions with two spin states with momentum between p and δp must satisfy the inequality

$$N(p)\delta p \leq (8\pi V p^2 / h^3)\delta p. \tag{A3.1}$$

The Maxwellian distribution written in terms of momentum using $p = mv$ and again using spherical polar coordinates in momentum space and considering a volume V in ordinary space is

$$N(p)\delta p = 4\pi n V (1/2\pi mkT)^{3/2} \exp(-p^2/2mkT)p^2 \delta p. \tag{A3.2}$$

233

If the Maxwellian distribution violates (A3.1), it will first do so at $p = 0$. This happens if

$$n^2 > 32\pi^3 m^3 k^3 T^3 / h^6. \tag{A3.3}$$

For a sufficiently high particle density n, the Maxwellian distribution violates the Pauli exclusion principle at low momenta. Some particles must as a result occupy higher momentum states than predicted by a Maxwellian distribution. Higher momentum implies higher kinetic energy and as a result the pressure is higher than the ideal classical value. Because of the m^3 on the right hand side of (A3.3), a much higher density is required for protons and neutrons than for electrons. A gas in which (A3.3) is satisfied is called a *degenerate gas*. In ordinary stars no particles but electrons are ever degenerate. In neutron stars, neutrons are degenerate. If a star can cool down and die, its gas must inevitably become degenerate.

The way in which a Maxwellian distribution is modified as the density increases at fixed temperature is shown in fig. 96. Eventually, to a good approximation, a situation is reached in which all states up to a critical momentum p_0 are occupied and all states beyond p_0 are empty. This is called a *completely degenerate gas* and it is strictly what happens at zero temperature. In such a gas I can write

$$\begin{aligned} N(p) &= 8\pi V p^2 / h^3, \quad p \le p_0 \\ N(p) &= 0. \quad\quad\quad p > p_0 \end{aligned} \Bigg\}. \tag{A3.4}$$

The total number of particles in volume V is N, given by

$$N = (8\pi V/h^3) \int_0^{p_0} p^2 \, dp$$

$$= 8\pi V p_0^3 / 3h^3. \tag{A3.5}$$

The pressure of a gas, P, is the mean rate of transport of momentum across a surface of unit area in the gas. This is

$$P = \int_0^\infty dp \int_0^{2\pi} d\varphi \int_0^\pi \sin\theta \, d\theta (N(p)/4\pi V) v_p \cos\theta p \cos\theta$$

$$= \frac{1}{3} \int_0^\infty (N(p)/V) p v_p \, dp, \tag{A3.6}$$

where v_p is the velocity of a particle with momentum p. The expression (A3.6) is constructed as follows. $N(p) dp / 4\pi V$ is the number of particles in unit volume with direction of motion in unit solid angle in momentum range dp. If they are moving at an angle θ to the surface considered the factor $v_p \cos\theta$ gives the rate at which they cross the surface and $p \cos\theta$ is the momentum carried across by each particle. The integral over all solid angle and p then gives the pressure.

In evaluating (A3.6) with the distribution (A3.4), I cannot simply use the relation $p = mv_p$ because at high density it is possible that $p_0 > mc$. As v_p must be

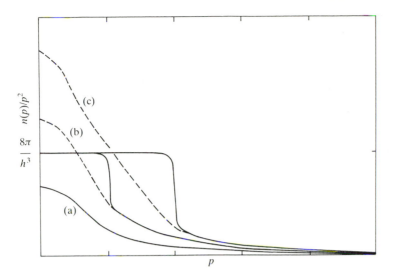

Fig. 96. The effect of the Pauli exclusion principle on the momentum distribution in a non-degenerate gas. (a) is the distribution in a non-degenerate gas. In (b) and (c) the dashed parts of the non-degenerate distribution above the critical value $8\pi/h^3$ are transferred to higher momentum as indicated by the solid curves.

less than c, I must use the relation between p and v_p given by the special theory of relativity. This is

$$v_p = (p/m)(1 + p^2/m^2c^2)^{-1/2}. \tag{A3.7}$$

The pressure of a completely degenerate gas can now be written

$$P = \frac{8\pi}{3mh^3}\int_0^{p_0} \frac{p^4\,dp}{(1 + p^2/m^2c^2)^{1/2}}. \tag{A3.8}$$

If I introduce

$$p_0/mc = x, \tag{A3.9}$$

it is possible to evaluate (A3.8) to give

$$P = (\pi m^4c^5/3h^3)[x(2x^2 - 3)(x^2+1)^{1/2} + 3\sinh^{-1} x], \tag{A3.10}$$

but I shall not give the derivation. The particle number density, n, can be written

$$n = N/V = (8\pi m^3c^3/3h^3)x^3. \tag{A3.11}$$

Equations (A3.10) and (A3.11) provide an expression for P in terms of n.

It is possible to obtain expressions for P valid either when $p_0 \ll mc$ or $p_0 \gg mc$ without using (A3.10), although the general expression can be used to check that the results are correct. In the first case p^2/m^2c^2 can be neglected in the integrand of (A3.8) and the integral gives

$$P = \frac{8\pi p_0^5}{15mh^3} = \frac{1}{20}\left(\frac{3}{\pi}\right)^{2/3}\frac{h^2}{m}n^{5/3},$$ (A3.12)

using (A3.11). When $p_0 \gg mc$, I neglect 1 in comparison with p^2/m^2c^2 in the integrand in (A3.8) and obtain

$$P = \frac{2\pi c p_0^4}{3h^3} = \frac{1}{8}\left(\frac{3}{\pi}\right)^{1/3}hcn^{4/3},$$ (A3.13)

again using (A3.12).

In the case of electron degeneracy in a fully ionised gas, there is a simple relationship between electron density, n_e, and mass density, ρ, if the chemical composition is specified. For each mass m_H of hydrogen there is one electron and to a very close approximation for helium and heavier elements there is half an electron for each m_H. Thus, if the hydrogen mass fraction is X,

$$n_e = \frac{\rho X}{m_H} + \frac{\rho(1-X)}{2m_H} = \frac{\rho(1+X)}{2m_H}.$$ (A3.14)

Insertion of (A3.14) into (A3.12) and (A3.13) gives the expressions for the pressure stated in Chapter 4 and used in Chapter 10. For a neutron star, the corresponding relation (neglecting the small fraction of protons and electrons also present and also the mass difference between the neutron and the hydrogen atom) is

$$n_n = \rho/m_H.$$ (A3.15)

The actual pressure in a neutron star is not however given by (A3.10) or by the limiting forms (A3.12), (A3.13) because at very high densities a repulsive force acting between neutrons also influences how tightly they can be packed and, as a result, their actual pressure is higher than the simple degeneracy value. The similar electrostatic repulsion between electrons is relatively less important as the density for electron degeneracy is much lower.

In many stars electron degeneracy is partial rather than complete. In that case there is no longer an analytical expression for the electron pressure. I will simply state the ingredients that go into the calculation of P. (A3.4) and (A3.5) must be replaced by

$$N(P)\,dp = (8\pi V p^2/h^3)\,dp \left/ \left[\exp\left(-\eta + \frac{E}{kT}\right) + 1\right]\right.,$$ (A3.16)

$$N = (8\pi V/h^3)\int_0^\infty \frac{p^2\,dp}{\left[\exp\left(-\eta + \frac{E}{kT}\right) + 1\right]}.$$ (A3.17)

Here E is the kinetic energy of an electron, which, according to special relativity, is given by

$$E = m_e c^2\left[(1 + p^2/m_e^2c^2)^{1/2} - 1\right]$$ (A3.18)

and η is determined by requiring $n_e \equiv N/V$ and T have specified values. If $\eta \rightarrow -\infty$, (A3.16) approaches the relativistic version of the Maxwellian distribution. If $\eta \rightarrow +\infty$, we recover the state of complete degeneracy. The pressure of a partially degenerate gas lies between the predictions of the ideal classical gas law and the complete degeneracy formula for given ρ and T.

Suggestions for further reading

A good (recently reprinted) book at a more elementary level is:

R. Kippenhahn, *100 Billion Suns*: *The Birth, Life and Death of the Stars*, Princeton University Press.

The best recent book which is generally at a slightly more advanced level than the present book is:

E. Böhm-Vitense, *Introduction to Stellar Astrophysics, Vol. 3 Stellar Structure and Evolution*, Cambridge University Press.

This is the final volume of a three-volume work on the properties of stars the other two being:

E. Böhm-Vitense, *Introduction to Stellar Astrophysics, Vol. 1 Basic Stellar Observations and Data*, Cambridge University Press.

E. Böhm-Vitense, *Introduction to Stellar Astrophysics, Vol. 2 Stellar Atmospheres*, Cambridge University Press.

A newly published elementary text on the physics of stellar interiors, which supplements the discussion of Chapter 4, is:

A.C. Phillips, *The Physics of Stars*, Wiley.

The best recent more advanced text on stellar evolution is:

R. Kippenhahn and A. Weigert, *Stellar Structure and Evolution*, Springer-Verlag.

An excellent older textbook, now republished in reprint form is:

D.D. Clayton, *Principles of Stellar Evolution and Nucleosynthesis*, Chicago University Press.

The classic texts on stellar structure (also now published in reprint form) are:

S. Chandrasekhar, *An Introduction to the Study of Stellar Structure*, Dover.

A.S. Eddington, *The Internal Constitution of the Stars*, Dover.

M. Schwarzschild, *Structure and Evolution of the Stars*, Dover.

Eddington's text was the original one on the subject. Most of its details are now wrong but many ideas are still correct. Chandrasekhar contains a large amount of mathematical detail. Both of these books antedate the development of the modern computer. Schwarzschild's book published in 1958 contains the first detailed discussion of extensive computations.

Index